T0281643

A Streetcar to Subduction
and Other Plate Tectonic Trips
by Public Transport
in San Francisco

Revised Edition

Clyde Wahrhaftig

Published by
American Geophysical Union

Streetcar to Subduction
Revised Edition

ISBN: 978-0-7590-234-0

Copyright 1984 American Geophysical Union
2000 Florida Avenue, N.W., Washington, D.C. 20009

Contents

Introduction

It is hard to be unaware of the earth in San Francisco. Built on rocky hills, the city is surrounded on three sides by bay and ocean that can be seen from nearly everywhere within it. Precipitous cliffs face the city from across the Golden Gate, and the skyline to the north, east, and south is dominated by mountains. Occasional tremors from the San Andreas and related faults nearby remind us that the earth here is active.

Until recently the rocks so abundantly exposed in San Francisco baffled geologists. Jumbled together without apparent order and lacking visible fossils, they defied explanation. The theory of plate tectonics has changed all that. We now have an explanation for the origin of the rocks of San Francisco, although it is anything but simple.

The theory of plate tectonics grew up in the pages of publications of the American Geophysical Union (AGU). This society holds its Fall Annual Meeting in San Francisco, early in December. In 1978, when Allan Cox was its president, I offered to prepare a guidebook to the plate tectonic outcrops of the city so that scientists attending the meetings could, in their off hours, go out and touch the physical embodiment of what they were discussing. The offer was accepted, and a guidebook with the same title as this one was published in a limited edition for free distribution to the participants of the 1979 meeting.

As part of the AGU contribution to the Decade of Geology—to increase public understanding of the earth—it was decided that when the guidebook was enlarged it should also be rewritten with less arcane jargon so that it could be used by nonscientists also. A brief review of the theory of plate tectonics as it applies to the rocks of San Francisco was prepared as an introduction. Those unacquainted with geology are urged to read it before making the trips. Some of the explanatory material is in individual trips: for example, on the origin and recognition of turbidity current deposits in Trip 3, on the metamorphic rocks here in Trip 6, and on the Great Valley Sequence in Trip 7.

It was not possible to eliminate all scientific terms, which are, after all, shorthand for conveying complicated concepts in a precise way. Therefore a glossary has been added in which technical terms and words used in a specialized way are defined or, if defined in the text, appropriately referenced. Definitions of other troubling words may be found in *The Dictionary of Geological Terms*, revised edition, prepared under the direction of the American Geological Institute and published by Anchor Press (1976), or *The Penguin Dictionary of Geology*, by D. G. A. Whitten and J. R. V. Brooks, published by Penguin Books, Ltd. (1972). An understanding of minerals and rocks can be obtained in the *Golden Field Guide: Rocks and Minerals*, by Charles A. Sorrell, illustrated by George F. Sandström, 280 pp., published by Western Publishing Co., Racine, Wisconsin (1973).

San Francisco has a fine public transportation system via which these technical features are easily accessible. I used it to prepare this guidebook, and if you use it for these trips, you will save money, energy (both animate and inanimate), and mental anguish.

Not all outcrops of tectonic significance are covered. There are no trips to the magnificent exposures of complexly folded chert in Golden Gate Park, nor to the rocks of Land's End. I had hoped to include a trip to the rocks west of the San Andreas Fault that are a displaced slice of the Sierra Nevada, whose granitic rocks are yet another manifestation of subduction. Limits of time and the economics of book publishing precluded their inclusion.

To increase accuracy (or at least verisimilitude), most field sketches of rock outcrops were made on sheets of transparent plastic placed over color Xerox prints of 35 mm color transparencies. This technique, although stifling to creativity, eliminates the influence of preconceptions. Its application resulted in modifying many of my ideas about the origin of these rocks and their structures, and I recommend it to others.

Acknowledgments

This guide could not have been prepared without the previously published work on the local geology by J. Schlocker, M. G. Bonilla, D. H. Radbruch, and Salem Rice, nor without the enormous amount of mapping and analysis of the Franciscan here and elsewhere by Tanya Atwater, E. H. Bailey, M. C. Blake, R. G. Coleman, W. R. Dickinson, Warren Hamilton, K. J. Hsü, W. P. Irwin, D. L. Jones, J. C. Maxwell, B. L. Murchey, Richard Ohrbohm, Ben Page, Emile Pasagno, and many others, nor without the studies of pillow basalt by J. C. Moore. Papers that report on and summarize this work are included in a selected bibliography. The book edited by W. G. Ernst is an up-to-date review.

Much of this guide was written some 8560 km from the outcrop, in the Sedgwick Museum, University of Cambridge. I am grateful to E. R. Oxburgh, Chairman, and Margaret Johnston, administrator, both of the Department of Earth Sciences, for excellent facilities for writing and preparation of illustrations and to many friends there who helped me with advice and information on ophiolites and omphacites, blueschist and boudinage, and other such arcane subjects, especially Stuart Agrell, Michael Carpenter, Tim Druitt, Richard W. Hey, Anton T. Kearsley, Dan McKenzie, Allan Smith, R. S. J. Sparks, John G. Spray, and Nigel Woodcock, and Evan Leitch of the University of Sydney.

Closer to home, "George" Schlocker and "Doc" Bonilla told me of many of the outcrops. M. C. Blake and Schlocker provided information on Angel Island and read a draft of that trip. Benita L. Murchey provided information on radiolarian stratigraphy. Robert G. Coleman and James Court made detailed comments on the earlier published version. Adrienne Morgan, Joan Gabelman, and Michael Lee drafted many of the maps. The staff of AGU, especially Ceil Tischler, Bill Laney, Margaret Connelley, Gregg Forte, Eric Garrison, and Cherry Fenwick worked hard to make this book a success. Allan Cox gave encouragement and valuable advice throughout the preparation of this guide, and it is dedicated to him.

The usage of two terms may be controversial. Current geologic usage restricts *serpentine* to the minerals and uses *serpentinite* for the rock. I find the latter term unnecessarily polysyllabic, as well as abrasive on the ear, and have used *serpentine* throughout. I tried to make the meaning clear in the context. I have used *melange* as it was first used by *Greenly* [1919].

The Theory of Plate Tectonics

The last two decades have seen a revolution in our understanding of how the earth works. We now know that the crust and upper mantle of the earth, to depths of at least several hundred kilometers, are in circulating motion, much like the motion of water in a teakettle or a beaker heated from below.

Along certain narrow belts beneath the oceans, called spreading zones, hot material from deep within the earth rises to the surface to form new ocean floor; this new material moves away from the spreading zones to make room for still more ocean floor. Elsewhere in the ocean, or along the margins of some continents, the ocean floor converges with parts of the earth's surface moving away from other spreading zones. There, the side of the convergence zone that is denser (that is, heavier) plunges back into the mantle. The downward plunging of part of the crust into the mantle is called subduction, and the zones along which it occurs are called subduction zones.

The ocean floor moves from the spreading zones to the subduction zones at rates of 2 to 20 cm per year. Snails are far faster; but at these rates, in 20 million years (m.y.) it would travel 400 to 4000 kilometers (km). The force that drives this motion seems ultimately to be the force of gravity, aided by heat from the interior of the earth. Most matter expands when it is heated, and its density therefore decreases. Hot matter from the interior of the earth is therefore less dense than the cooler material of the same composition that may be on either side of it; it rises because of its buoyancy, much as hot water rises from the bottom of a heated vessel.

These spreading zones of hot expanded material mark rises or ridges that run more or less along the center of the world's oceans. The oceanic crust cools and contracts as it moves away from the spreading zones, and therefore its surface, which is the ocean floor, gradually sinks. The oceans, in consequence, deepen away from the spreading zones. The general model is illustrated in Figure 1.

Most subduction zones are (or were) along the margins of continents, which rise above sea level because the continental crust is made of intrinsically less dense material than the oceanic crust. At the subduction zone the oceanic crust is dragged down by gravity (that is, subducted) beneath the less dense continental crust. Deep oceanic trenches mark most active subduction zones.

When it was discovered that the ocean floor is constantly being created at spreading zones, it was concluded that, except for being formed at spreading zones and consumed at subduction zones, the crust behaves as though it were an assemblage of rigid plates. Hence, this new explanation of the workings of the earth was called plate tectonics. Tectonics, derived from the ancient Greek word for building a house, is a geological term that includes the folding and faulting processes that built mountains.

In addition to spreading and subduction zones there is another kind of boundary between plates where they simply slide past each other. This kind of boundary is called a transform fault. Transform faults arise through two different mechanisms: (1) Where the initial spreading zone has formed not exactly perpendicular to the direction of spreading, it tends to get reorganized into a series of segments that are at right angles to the spreading direction and are joined by transform faults parallel to the spreading direction. (2) Both the spreading zones and the subduction zones can and do migrate and may ultimately converge. Where they do converge, the plate between them is consumed, and the motion along the single plate boundary that results is a motion between the two outer plates. Since spread-

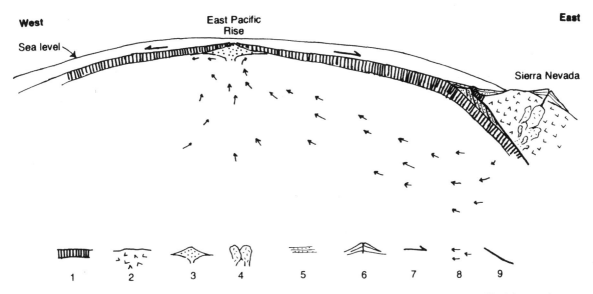

West

East Pacific Rise

East

Sea level

Sierra Nevada

1   2   3   4   5   6   7   8   9

Fig. 1. Diagrammatic, hypothetical cross section through spreading zone (East Pacific Rise), subduction zone (coastal California), and intervening oceanic plate (Farallon Plate) in Cretaceous time: **1,** oceanic crust; **2,** continental crust; **3,** gabbroic magma along spreading zone; **4,** dioritic-to-granitic magma beneath volcanic arc over subduction zone; **5,** turbidite sediments; **6,** volcanos over subduction zone; **7,** arrow indicating rapid motion of lithosphere away from spreading zone; **8,** arrows indicating slow diffuse motion within the upper mantle toward the spreading zone; **9,** subduction-zone thrust faults.

ing and subduction cancel each other when the two boundaries merge, the remaining motion is likely to be a lateral motion of one of the remaining plates past the other. This appears to have happened when part of the East Pacific Rise encountered the subduction zone along the western margin of North America: The result was the creation of the San Andreas Fault, the locus of sliding of the Pacific Plate past the North American Plate and one of the first transform faults to be recognized as such [*Wilson,* 1965].

In the broad outlines of plate tectonic theory, the earth's surface is thought to consist of about 10 or 12 major plates. The boundaries between the plates are where most earthquakes and volcanic activity, and hence most mountain building, are taking place. In the original conception of plate tectonics the interiors of the plates, with few exceptions, were thought to be relatively quiet. We now know that the picture is not that simple. Some plates have deformed internally, and plate boundaries, rather than being sharp lines, are somewhat diffuse bands with many microplates, like small slivers and blocks, sliding laterally and rotating between the major plates along some of the boundaries. The pattern of currently active plate boundaries is shown in Figure 2.

Within and near San Francisco are manifestations of all three kinds of plate boundaries. As Figure 6 shows, San Francisco lies next to a transform plate boundary, the San Andreas Fault. One of the com-

mon rocks of San Francisco and adjacent Marin County is pillow basalt, which was probably erupted along a spreading zone to form an ancient ocean floor. The bedrock beneath San Francisco and much of Marin County as well as much of the California Coast Ranges, called the Franciscan Formation (or Franciscan Assemblage in some reports), is ocean-floor basalt and the sediments deposited on it, all of which were stacked against and beneath the western edge of the North American continent as a result of the subduction beneath North America of an ancient Pacific oceanic crust. For this reason other reports and maps call it the Franciscan subduction complex, or Franciscan Complex for short. In this guidebook, to avoid taking sides in a purely semantic controversy, it will simply be called the Franciscan.

In order to understand these rocks in the context of plate tectonics we have to consider briefly what happens as new oceanic crust forms at a spreading zone, what is deposited on that crust as it migrates toward a subduction zone, and finally, what happens to it during subduction. These are considered in the next three sections.

The Origin of the Ocean Floor

The new ocean floor at the spreading zone consists of dark basalt lava that is erupted onto the

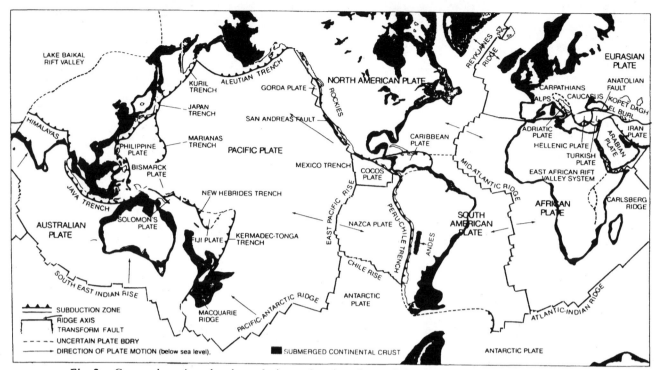

Fig. 2. Currently active plate boundaries and major plates of the earth's lithosphere (after *Dewey* [1972], copyright 1973 by Scientific American, Inc., all rights reserved; published with permission).

ocean floor through a crack that opens periodically along each segment of the spreading zone. The eruption of the basalt is not continuous but occurs whenever the crack pulls apart far enough to fracture the recently frozen lava. The basalt that erupts is not the material that rose from the depths of the mantle, which is thought to consist mainly of heavy minerals rich in iron, magnesium, and silicon; rather the erupting basalt is the low-melting fraction of the mantle material, richer in calcium, sodium, and aluminum, and containing traces of potassium and other elements.

The most characteristic form taken by the submarine basalt is an accumulation of small bulbous masses called pillows. Cold seawater chills the lava as it erupts, forming a thin crust (a few millimeters thick) of solid basaltic glass over a bulb of liquid lava. Lava continues to pour into the bulb, enlarging it and keeping the glassy crust broken along one or two continuously open cracks along which the surface of the bulb widens and chills to form additional glassy crust (a small-scale analog of seafloor spreading). The bulbous mass of lava is the pillow, with a convex upper surface. It is usually molded to the tops of older pillows and has a keel projecting downward along their junctions (see Figures 3 and 35). When the pillow reaches a meter or so in diameter, its surface grows so slowly that the crack freezes over and the pillow stops growing. The lava breaks out elsewhere to form a new pillow.

When a local eruption ceases, the lava left in the feeder crack, at depth, congeals to form a thin, vertical sheet called a dike. Repeated opening of these deep cracks along the weak hot rock of the youngest dike results in a multitude of dikes that have one side with a chilled border against older rock and the other side showing younger rock chilled against the dike. Because the lava in the dikes cools more slowly than on the seafloor, it crystallizes to a somewhat coarser rock called diabase.

The dikes pass downward into a reservoir of magma (that is, molten lava that has not reached the surface). As the sides of the magma reservoir move away from the spreading zone, the magma slowly cools and ultimately crystallizes to a dark, coarse-grained rock called gabbro, which has the same composition as basalt. However, the mineral grains that first crystallize are olivine and pyroxene, both denser than the magma. These settle to the floor of the magma reservoir, where they accumulate as layers of rock consisting solely of these two minerals. Rocks consisting solely of such dark minerals are called ultramafic rocks, a term derived from the fact that they contain a superabundance of magnesium and iron (whose latin name is ferrum).

The lowest layer in the sequence of rocks beneath the ocean floor is the unmelted fraction of the mantle, which also consists almost entirely of olivine and pyroxene and hence is an ultramafic rock. Thus the complete sequence of rocks beneath the young

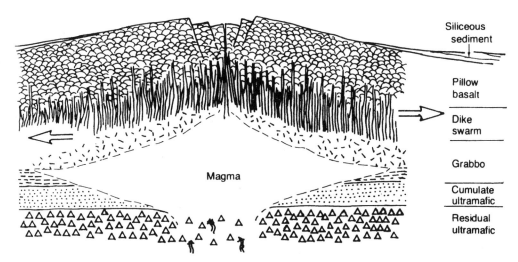

Sea level

Siliceous
sediment

Pillow
basalt

Dike
swarm

Grabbo

Cumulate
ultramafic

Residual
ultramafic

Magma

Fig. 3. Diagram showing the genesis of oceanic crust at a spreading zone (based on *Coleman* [1977], p. 36).

ocean floor consists of pillow basalt at the top that is intruded at depth by a layer of dikes, beneath which is gabbro, and beneath the gabbro are two layers of ultramafic rock (Figure 3). The rare occurrences of such ocean floor sequences found on land are called ophiolites.

The lava erupted on the ocean floor has numerous pores, cracks, and cavities, and as the dikes beneath cool they contract to form more cracks. Seawater circulating through these openings is heated by the hot rock and reacts with the basalt and underlying diabase to form minerals that contain water, to replace some of the basalt with jasper, and to concentrate some of the rare valuable elements, such as copper, into ore deposits. The recently discovered hot springs along the Galapagos Rise and at the mouth of the Gulf of California, with their exotic colonies of undersea life, are where such water, with dissolved sulfur, returns to the ocean, and the fantastic life around them gets its energy via a food chain based on the oxidation of the sulfur by bacteria.

The Seafloor Sediments

In the sunlit surface waters of the sea, myriad minute, one-celled organisms secrete shells of lime (calcium carbonate) or of opal (a form of silica). When these organisms die their shells sink slowly through the water, forming what Rachel Carson in *The Sea Around Us* called "the long snowfall." The lime-secreting organisms include foraminifera and various one-celled algae; those secreting opal in-

clude radiolarians and diatoms. Mixed with these minute shells is a fine mineral dust, usually red, blown from the desert areas of the continents.

In most of the oceans the carbonate shells dissolve before they reach the ocean floor; the radiolarian shells remain and, together with the mineral dust, form the red radiolarian ooze of the deep ocean floors. In the tropics the carbonate shells are more abundant; more of them reach the ocean floor and, together with the mineral dust, form a pink to red limy mud.

The rate of accumulation of the radiolarian ooze is extremely slow, so that by the time the ocean floor has migrated a few thousand kilometers from the spreading zone, only a few tens to a few hundred meters of ooze have accumulated. Shortly after its accumulation, the siliceous ooze apparently segregates into thin layers of nearly pure silica, which hardens into chert, separated by even thinner layers of shale. We call this thinly layered rock, whose outcrops are spectacularly striped, ribbon chert.

As the ocean floor approaches a continental margin or volcanic island arc such as the Aleutians, where it will encounter a subduction zone, the ooze and limy mud are covered, first, with mud that has floated far out to sea before settling and, on top of that, with deposits of the giant submarine sand- and mudflows, called turbidity currents, that race down the submarine canyons that score the continental slopes, where they spread out as giant submarine fans on the deep ocean floor near the continents. The source of the sediment in these currents may ultimately have been sand eroded off the continents by rivers or along the shore by waves, or volcanic ash from eruptions along the volcanic island arc or chains of volcanos on land. These massive turbidity current deposits, called turbidites, accumulate to sequences thousands of meters thick, partially or completely filling the trench along the down-going side of the subduction zone (Figure 1).

What Goes on During Subduction

Not all the material beneath the ocean floor is subducted into the mantle. Subduction mainly affects the heavier rocks. Some of the light sedimentary material and even some of the pillow basalt is scraped off the down-going slab of oceanic crust to be stacked against or immediately beneath the base of the upper plate of the subduction zone. This base is usually a great fault, dipping 10°–30° beneath the upper plate, along which the subducted oceanic crust slides into the mantle. Some of the sedimentary and volcanic rock is carried to great depths, where it is altered into rather unusual metamorphic rocks, such as blueschist and eclogite; some of these, along with fragments of ultramafic rock from the mantle, are squeezed as blocks back upward toward the surface. Much of the sedimentary rock stacked beneath the upper plate acts as a lubricant for the sliding and is crushed and mixed as in a giant mill. The mixing has stirred the blocks of the unusual metamorphic rocks, the ultramafic rock, and the blocks of chert, basalt, gabbro, and sandstone into the thoroughly mixed and crushed material, creating a gigantic lithologic "plum pudding" we call melange. In San Francisco and Marin counties we see all gradations between well-preserved volcanic and sedimentary ocean floor rocks and melange in which the rocks have been crushed beyond recognition.

The subducted material is heated as it descends and gives off its charge of volatile substances picked up by reaction with seawater—such substances as water, chlorine, sulfur, and carbon dioxide. These rise through the overlying plate, causing parts of that plate to melt partially and stirring already molten material there. This molten material (magma) rises, in part, to erupt at the surface as a chain of volcanos. If the volcanos grow upward from the ocean floor, they form an island arc, such as the Aleutians. Similar lines of volcanos grow along the continental sides of subduction zones: examples are the Cascade Mountains and the great volcanos of the Andes. The material erupted from these continental volcanic arcs is usually andesite and dacite.

Much of the magma generated at subduction zones does not reach the surface but forms great

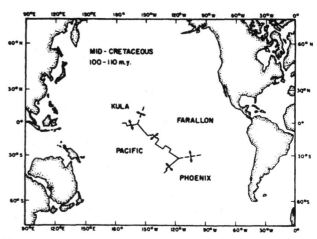

Fig. 4. Reconstruction of the Pacific Ocean in the middle of the Cretaceous (from *Alvarez et al.* [1980]).

bubbles of magma within the crust beneath the volcanos. This magma crystallizes to form bodies of coarse-grained plutonic rocks, e.g., diorite, granodiorite, and granite, such as we see in the Sierra Nevada. Examples can also be seen in Point Reyes National Seashore and along the coast south of San Francisco and west of the San Andreas Fault. Much of the material of the sedimentary accumulations in the subduction zone trenches is the sand formed by the disintegration, through weathering, of these plutonic rocks, after the volcanic rocks above them have been eroded away. Such sandy detritus makes up much of the Franciscan and the Great Valley Sequence to the east, described in Trip 7.

A hypothetical section through the subduction zone that probably existed along California 100 m.y. ago is shown in Figure 45b. The infrared position of the spreading zone at about that time is shown in Figure 4.

The rim of the Pacific Ocean is almost entirely encircled by subduction zones (see Figure 2), which are marked by lines of volcanos, such as the Andean, Central American, Cascade, and Aleutian volcanos. The subduction belt begins in New Zealand, sweeps through Tonga, Samoa, the New Hebrides and Solomon islands, north through the Marianas, Japan, the Kuriles, and Kamchatka, thence east along the Aleutians, and south along most of the American coastline. The two major gaps in this belt of subduction zones are along southeastern Alaska and British Columbia and from Cape Mendocino to the mouth of the Gulf of California. At these two gaps the Pacific Ocean spreading zone (the East Pacific Rise), which had been migrating eastward toward North America, ultimately encountered the subduction zone that was drifting west from the Mid-Atlantic Ridge. Where this happened, the intervening plate, called the Farallon Plate, disappeared, and the resulting motion of the two outer plates is horizontal, i.e., the Pacific Plate slides northwesterly past the North American Plate, along the San Andreas and its associated faults.

Hence subduction no longer occurs at San Francisco, and all one can see here are its products. If you wish to experience subduction in action, you must go to Alaska or Washington state. Recent events in those areas, the Great Alaskan Earthquake of 1964 and the eruption of Mt. St. Helens in 1980, attest to the power of a geologic process we have just begun to understand.

The Franciscan in San Francisco

The Franciscan in San Francisco consists predominantly of five rock types: (1) dark, dirty sandstone, called greywacke, which is interbedded with lesser amounts of (2) dark grey claystone and shale; (3) thin-bedded ribbon chert, generally red, locally with abundant radiolarians: (4) somewhat altered submarine pillow basalt, commonly called greenstone; and at a few localities, (5) schist, some of it glaucophane-bearing and indicative of high-pressure and low-temperature metamorphism.

What the Rocks Look Like

The sandstone, normally dark grey and quite hard, typically weathers light brown to buff. This weathering process breaks it into small fragments, and it ultimately becomes friable. Weathering can extend to depths of over 20 m. Similarly, the basalt, which is dark green to black and very hard when fresh, weathers to a friable orange mixture of clay minerals and ferric oxides. Weathered sandstone can be recognized as sandstone by its yellowish color, its clastic texture, the presence of grains of quartz and chert, and by the fact that it breaks into sharply angular fragments with planar surfaces. Recognizing weathered greenstone is more difficult: it is orange to reddish-brown and fractures into fragments with curved surfaces that are usually stained brown to black. The texture of basalt—an interlocking network of thin, light-colored, flat, euhedral plagioclase platelets with dark pyroxene filling the interstices known as ophitic texture—persists through weathering and can be seen with a high-power hand lens on a freshly broken unstained surface, even though the feldspar has been altered to chalky white clay minerals and the pyroxene to an orange or brown mixture of clay and iron oxides.

The chert weathers little but breaks into small, hard, roughly rectangular fragments with dull, smooth surfaces. The radiolarians can be seen as translucent dark grey specks if the chert is wetted and examined in bright sunlight with a powerful hand lens.

Intercalated with the Franciscan are belts of serpentine, the state rock of California, noted for its propensity to form landslides and, incidentally, formed by the hydration of ultramafic rocks from the mantle. Practically all ultramafic rocks exposed near San Francisco have been converted to serpentine. Exposures of serpentine are typically pale green to greenish-gray, locally black. Like the chert, it does not weather readily, and its outcrops in road cuts and sea cliffs usually expose unweathered rock.

The serpentine occurs in two "structural styles": (1) as somewhat rounded blocks of dense, usually dark-green, massive serpentine embedded in (2) intensely foliated pale-green serpentine that splits readily along greasy-feeling mirror-smooth foliation surfaces that wrap around the massive blocks. The latter type is frequently called "slickentite." The serpentine in the massive blocks can occur in three forms: (a) as a dense aggregate of exceedingly fine crystals—so fine as to be invisible—a replacement of olivine; (b) as isolated bodies 2–5 mm across that are embedded in (a) and that have an excellent parting resembling cleavage and that are replacements of pyroxene; this particular form of serpentine is called "bastite"; (c) closely spaced networks of 1–5-mm-thick veinlets of asbestos serpentine, which give a "crazed-china" pattern to some exposures. The bastite grains may be in layers or exhibit a vague planar parallelism, possibly a relict of tectonic deformation of the ultramafic rock when it was still in the mantle.

Origin of the Franciscan

Plate tectonics has greatly clarified the origin of the Franciscan. Once thought to be a shallow marine, or even terrestrial, accumulation [Davis, 1918], it is now recognized as an ocean floor assemblage that has suffered the effects of subduction. Some of the basalt is the original volcanic crust formed at a mid-ocean spreading zone, and some of it is volcanic accumulations that formed on the seafloor as the latter migrated from a spreading zone to a subduction zone. The chert is derived from the radiolarian ooze that rained down on the young ocean floor and volcanic seamounts. The much more voluminous sandstone and shale accumulated as turbidite sequences as the ocean floor approached the continent.

Fossils found in the Franciscan in San Francisco and along the Golden Gate range in age from mid-

Jurassic to mid-Cretaceous, and include radiolarians from the chert and ammonites and pelecypods rarely found in the sandstone [*Murchey*, 1980; *Wright*, 1974; *Armstrong and Gallagher*, 1977].

The route by which Franciscan rocks traveled to San Francisco is still under contention. One view is that they were accreted to North America by subduction that occurred at their present latitude. The location of the source of the ocean floor according to this view is shown in Figure 4. The other view is that they accreted to the south of here and moved to their present location by strike-slip motion along the continental margin, much as California west of the San Andreas Fault is migrating northward in relation to the rest of North America.

Accretion against and beneath the leading edge of North America was accompanied by intense deformation. While some of the Franciscan was accreted as blocks whose rocks, although folded and faulted, are still relatively intact, large volumes of the Franciscan were reduced to the intensely sheared paste with embedded exotic blocks we call melange. All gradations, from coherent Franciscan through "broken formation" (crushed sandstone and shale without exotic blocks) to melange, exist in and near the city.

Speculations on the Structure of the City

In San Francisco, five northwest-trending belts of distinctive rock assemblages (labeled I–V in Figure 5) are apparently the outcropping edges of northeast-dipping structural slabs. The southwesternmost slab (I), the San Bruno Mountain Block, consists of fairly well-bedded sandstone and shale exposed in and around San Bruno Mountain (Figure 6) and at the Cliff House (Figure 5). It contains little-deformed turbidite sequences, to be seen on Trip 4, and the sandstone has a significant amount of K-feldspar, which traditionally has been thought to indicate a probable Cretaceous age by correlation with the Great Valley Sequence to the east of the Coast Range Fault (See Trip 7). The San Bruno Mountain Block was possibly the latest block within the city to have been stacked against the continental margin by subduction.

The San Bruno Mountain Block is bordered on the northeast by a belt (II, Figure 5) of melange 0.6–1.8 km wide that extends from the landslides at Land's End, southeasterly beneath dune fields, through City College and Visitacion Valley, to Bayshore. The southern boundary of melange is along Geneva Avenue, and its northeast contact may be exposed on the south side of Candlestick Hill. In Visitacion Valley and John McLaren Park (Figure 7) are exotic blocks of sandstone, gabbro, basalt, ser-

pentine, and schist, which now form picturesque rock knobs, also to be seen on Trip 4.

Northeast of the melange belt is the Twin Peaks Block (III), characterized by chert and basalt but also with sandstone and shale. This block is cut by many zones of shearing, some of which are sheets of melange. This block includes most of Golden Gate Park, the central highlands of the city (Twin Peaks, Mt. Sutro, Mt. Davidson, and Diamond Heights), Bernal Heights, and Candlestick Hill. Trips 1 and 4a are mainly to exposures in this block. The Twin Peaks Block bears a close resemblance to the Southern Marin Block (see below), with which it may be coeval.

Northeast of the Twin Peaks Block a diagonal belt (IV) across the city is dominated by serpentine. This belt extends from Fort Point (the south abutment of Golden Gate Bridge, Trip 3) through the New Mint Building (Trip 1) and Potrero Hill to Hunter's Point. In some exposures, sheets of intensely deformed serpentine alternate with sheets of melange. Other exposures are of massive serpentine with large blocks of gabbro. Foliation and layering within this belt generally dip northeast.

The northeasternmost block, which will be called the Nob Hill Block (V, Figure 5), consists largely of sandstone and shale exposed at Fort Mason (Trip 2) and on Nob and Russian hills. The great sandstone cliff on the east side of Telegraph Hill (Trip 4c), apparently the quarried edge of a single enormous sand flow (J. Schlocker, oral communication, 1967), is its most spectacular exposure. Yerba Buena and Alcatraz islands may be part of this block. The Nob Hill Block appears to have been folded into a broad syncline whose hinge runs down Columbus Avenue, and the hinge of an anticline may lie beneath the bay, west of Yerba Buena Island. The sandstone of this block lacks significant K-feldspar [*Wright*, 1974].

The complicated structure of the Franciscan formed at great depth in the subduction zone. Several hundred, or more likely several thousand, meters of once overlying rocks have been eroded away in the 60 m.y. since the structures we see in the Franciscan were formed, and the present land surface is an irregular horizontal cut through the Franciscan. The Quaternary deposits shown on Figure 5 were laid down on that irregular surface.

The Southern Marin Block

The Southern Marin Block is an anomalous mass of coherent Franciscan that makes up the Golden Gate Headlands and extends for about 4.5 km north into Marin County (Figure 30). It is in fault contact with a great body of melange that makes up

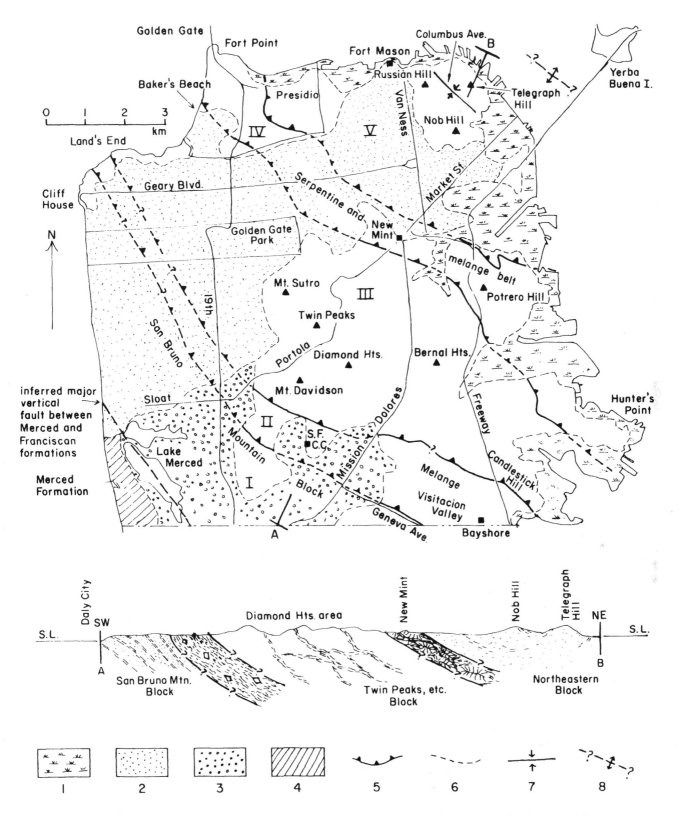

Fig. 5. Generalized geologic sketch map of San Francisco (generalized and modified from *Blake et al.* [1974]) with diagrammatic geologic section: **1,** filled land; **2,** area covered for most part by Holocene dune sand; **3,** area covered for most part by Pleistocene Colma Formation (dune sand and lagoonal and alluvial deposits); **4,** area underlain by the Plio-Pleistocene Merced Formation (Blank areas have Franciscan rocks at the surface.); **5,** boundary between major blocks in the Franciscan (dashed where covered by surficial deposits; blocks are indicated by Roman numerals; boundaries are probably thrust faults; black triangles on down-dip side of faults); **6,** contacts of surficial (Quaternary) deposits; **7,** synclinal hinge line; **8,** inferred anticlinal hinge line.

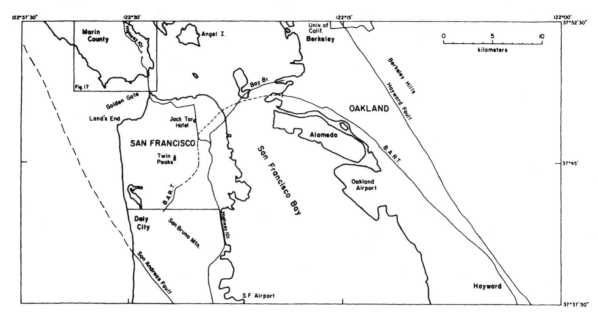

Fig. 6. Index map of San Francisco and vicinity.

most of Marin County east of the San Andreas Fault. This fault contact dips gently south beneath the eastern half of the Southern Marin Block and is nearly vertical farther west. The block is an assemblage of lenticular to tabular bodies of basalt, ribbon chert, sandstone, and shale, each a few to a few hundred meters thick and a few hundred meters to 3 km or more long.

The rocks, in general, strike northwest and dip southwest in the eastern part of the block and bend around to strike west and dip south in its western part. The main structure is thus a broad, shallow southwest-plunging syncline. Pillows in the basalt, and sedimentary structures such as cross-bedding and minor channeling in the sandstone, consistently indicate younger rocks to the southwest. Hence the whole assemblage is right side up. However, studies of the radiolarians in the chert by *Murchey* [1980; oral communication, 1982] have shown that the chert sequence in several of the bodies ranges over the same span of geologic time, from Jurassic to mid-Cretaceous. Thus, the block apparently consists of a repeated stacking of slabs, which consist of sandstone resting on chert resting on basalt—in other words, fragments of a Mesozoic ocean floor.

The contact beneath the Golden Gate between the Southern Marin Block and the Franciscan of San Francisco appears to be a profound tectonic dislocation, because an east-dipping terrane of serpentine, melange, and sandstone south of the Golden Gate (Trip 3) and a west-dipping terrane dominated by chert and basalt and lacking serpentine, north of the Golden Gate, strike directly toward each other. The nature of this dislocation, whether a vertical or nearly flat fault and whether strike-slip or dip-slip,

is unknown. Trip 5 is to spectacular exposures of pillow basalt and tightly folded chert in this block.

Trip 1. A Streetcar to Subduction

This is a streetcar ride to see parts of San Francisco's late Mesozoic subduction zone complex, the Franciscan. It starts where pillow basalt, chert, and sandstone are in a deceptively simple and orderly relationship, then takes you to successively more complicated and tectonically disturbed outcrops, and ends in a mass of serpentine. The trip has several segments, each of which can be taken separately. The directions to the separate parts (labeled B, C, D, etc.) are given in the text at the appropriate places.

Take the Muni Number 47 trolleybus or Number 42 (red arrow) motor bus southbound on Van Ness Avenue (get a transfer). At Van Ness and Market (Figure 7), enter the Muni Metro Station and take the "J" (Church) streetcar outbound—the streetcar to subduction. Where the J car emerges from a subway into daylight, (Figure 8) is a large grey building on the right. It sits on a trimmed-off stump of hill and is the New Mint Building. (It was built in 1937 to be impregnable to burglars and was broken into by two high-school students on a lark 2 weeks after it was dedicated.) It sits on serpentine, part of the serpentine belt that crosses San Francisco from Fort Point to Hunter's Point (Figure 5, IV). Even from the streetcar, it is possible to see that the shearing and veining dip northeast. The exposure on the

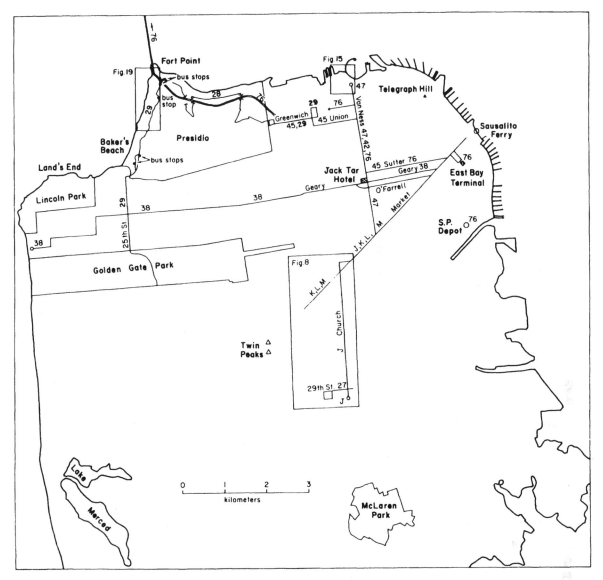

Fig. 7. Outline map of San Francisco, showing Muni lines to Trips 1, 2, and 3 and locations of Figures 8, 15, and 19.

next block, which is more accessible, more photogenic, and more instructive, is the last stop on this tectonic trip.

The streetcar turns left or almost due south, onto Church Street. Four blocks south of Market the tracks leave the street, enjoying their own right-of-way through Dolores Park and around a steep hill beyond, then rejoin Church at 22nd. The energetic should stay in the car to the end of the line. The lazy can transfer at 29th Street (beside the gothic granite pile of St. Paul's Church) to the Number 27 bus uphill to the west. Ask the bus driver to let you off at Castro and 30th and you will be beside the first pillow basalt outcrop.

Those who ride to the end of the line should walk west on 30th three long blocks to the little hillside open space—Billy Goat Hill, a natural-area park set aside by the city about 7 years ago. The three up-

permost layers of typical oceanic crust—pillow basalt, siliceous ooze (now radiolarian chert) and sandy turbidite (now greywacke)—are exposed here (Figure 8, A; Figure 9) in microcosm. The road cut cliffs at (1), Figure 9, along 30th Street at the foot of the hill, expose weathered pillow basalt. Deep weathering has altered the basalt from its hard, blackish-green condition to the soft, orange-brown material and has fractured it so thoroughly that the pillow outlines are hard to recognize. However, at (2), near the west end of the outcrop, pillows can be made out in cross section, and they indicate that the rock is right side up. Two steep paths up the hill have natural steps on the worn tops of pillows.

The contact between pillow basalt and overlying chert probably reaches road level at (3), where the chert appears to be only 3 m thick but is probably faulted against the basalt. The bushes growing on the chert outcrop are poison oak (*Rhus diversiloba*) and should be avoided. They can be recognized by the shiny, smooth-lobed leaves in triplets that turn

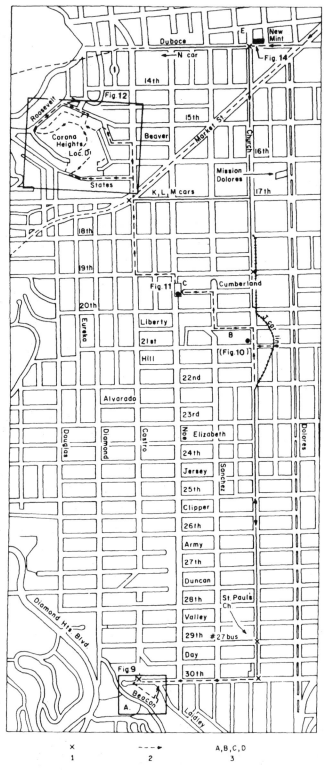

Fig. 8. Street map of part of San Francisco, showing destinations of Trip 1 and locations of Figures 9, 10, 11, 12, and 14. East-west width of this map is approximately 1500 m (5000 ft); **1**, streetcar or bus stop; **2**, path of walking Trip; **3**, locality referred to in text.

bright red in late summer and by the stiff, slightly orange stems when the leaves are off. Poison oak is present on nearly all these trips.

A trail leads diagonally uphill from (4), past grey quarry waste (relict of a large, crushed-rock quarry that once operated at the end of 30th Street), to join a contour trail to a flat bench on a large chert outcrop (5) at the nose of the hill. The red, radiolarian-bearing chert is in layers 2–10 cm thick that are interbedded with red shale partings about 5 mm thick. The beds are gently folded into open wave-like folds, with wavelengths of 1–2 m, plunging gently to the north. The chert is seamed by closely spaced quartz veinlets that are perpendicular to the bedding and that trend in three directions about 120° apart. The chert dips about 20° northwest.

Sandstone is exposed along the ridge crest to the southwest; the finest exposure is at (6), where the ridge broadens to an artificial flat at the level of Beacon Street. The sandstone is typical medium-grained greywacke, made up of angular sand-size fragments of quartz, feldspar, and volcanic rocks embedded in a clay-rich matrix. Thin interbeds of shale and fine sandstone in a shallow quarry on the southeast side, (7), indicate that the sandstone is nearly flat-lying.

The apparent simplicity of structure here is deceptive. With a little study in this neighborhood, you would soon recognize that these rocks are part of a highly deformed sequence. This locality may be either a fault-bounded sliver or the lower limb of an overturned syncline.

Return east on 30th Street to the terminus of the J streetcar. The high hill directly ahead, topped by a microwave repeater station, is Bernal Heights, with spectacular exposures of chert. Take the J car north to 22nd or 21st street. Walk north on Church or west on 21st (uphill in either case) to the northwest corner of Church and 21st (B, Figure 8). On the Church Street side of the apartment house before you, the building is supported on pillars to permit you to view the Franciscan bedrock (Figure 10.) Blocks of massive sandstone are interspersed with intricately branching seams of sheared shale that contain bodies of crushed greenstone, sandstone, and chert. A few blocks of solid greenstone are present. The apparent dip of shearing is roughly 15°– 20° north. This may be typical of an early stage in the production of melange matrix. The owner and architect are to be commended for preserving these valuable exposures. Please do not do anything, such as collect samples or trespass on the outcrop, that might cause them to cement it over. To start the trip at this exposure, get off the southbound streetcar on 21st Street and walk uphill.

Walk north down Church to Liberty. If you have to end the trip here, walk one block and catch the J

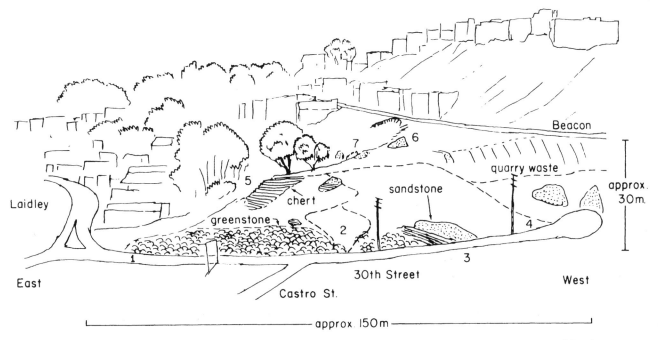

Fig. 9. View south to Billy Goat Hill, showing relationships of rock units, rock exposures, and localities mentioned in text. Sketched from a photograph. Dashed lines are paths.

car northbound at the southwest corner of Dolores Park, or walk diagonally across the park and north two blocks on Dolores Street to see Mission Dolores, original point of European settlement in San Francisco in 1776. Four blocks farther on Dolores brings you to Market, the New Mint, and the outcrops that end this trip.

If you're still feeling adventuresome, turn left (west) on the north side of Liberty to Sanchez, right (north) on Sanchez 1½ blocks to Cumberland, and left (west) on Cumberland to Noe. This is an elegant and architecturally interesting residential neighborhood, with fine views of Berkeley and Oakland across the bay. At the west end of Cumberland (point C, Figure 8), a flight of steps and a path descend over an outcrop of relatively unweathered greenstone (slightly metamorphosed basalt) to the north and thin-bedded shale and sandstone to the south.

Figure 11 is a sketch map of this locality and shows that the contact is probably a fault. Pillow structure in the greenstone is obscure, part of the rock may actually be intrusive into the remainder, but suggestions of pillow structure at the north end of the exposure, if correctly interpreted, indicate a dip of 45° north or northeast and younger greenstone to the north. This is an excellent place to see the color and characteristics of unweathered greenstone.

Walk north on Noe to 19th, left (west) on 19th to Castro, and right (north) on Castro. You are now in the heart of San Francisco's gay community. Don't get subduced. Cross Market (on west side of Castro or through subway station), and one short block beyond Market turn west on States Street. You can start the trip at Market and Castro by taking an outbound "L", "K", or "M" streetcar in the Muni Metro and getting off at Castro station.

The bedding in a few obscure outcrops of chert on the north side of States Street appears to dip 20°–40° southeast. Beyond about 100 m west of Castro, the Corona Heights Playground borders States Street on the north (D, Figure 8; Figure 12). This was once a brickyard that manufactured bricks from local clay. Partly buried remains of brick structures, piles of waste brick, and a brick pavement buried by what at first appears to be alluvium are exposed along the north side of the street. The brickyard caught fire during the 1906 earthquake, leading to rumors of volcanic eruption on Corona Heights. The quarry floors provide the flat areas on which the playgrounds, tennis courts, picnic areas, and the Josephine Randall Junior Museum are located. Their walls provide superb exposures of Franciscan rocks.

Take the first paved driveway to the right uphill to a small playground with tennis and basketball courts and restrooms. A greensward, level with this playground, extends east and north around the southeast corner of the hill and is backed by a cliff of chert. Pillow basalt forms the west end of the cliff and is deeply weathered, making the pillow structure obscure, but right-side-up pillows can be recognized at (1), Figure 12, at the sharp bend of the paved path leading to the level above. The pillow basalt may be a thin flow in the chert, for chert is

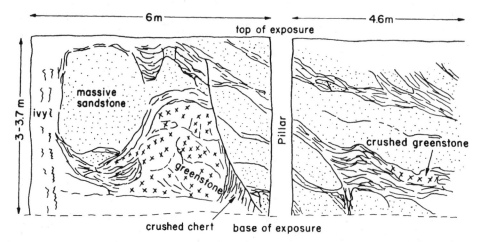

Fig. 10. Field sketch of exposure beneath apartment building on west side of Church Street, just north of 21st Street (locality B, Figure 8).

exposed below it on States Street and on the access driveway.

At (2), just east of where the top of the pillow basalt intersects the parking lot level, is a nearly isoclinal recumbent fold in chert with a hinge line trending N 55°E. A few feet above this fold is a 1-m-thick layer of deeply weathered basalt, apparently comformable with the chert. Relicts of the tiny interlocking crystals can be seen when a fragment is examined with a powerful hand lens (see section above entitled What the Rocks Look Like). Minor folds in thin-bedded chert are exposed at several places along the cliff to the east (3, 4). Their hinge lines trend N 50°–60°E, and their axial planes strike northerly and dip 10°–25° east. The minor folds have one long limb and one short limb. The short

limbs are immediately below the hinge lines about which the folds are convex to the northwest. This arrangement suggests that this is possibly the lower limb of a large recumbent syncline open to the west.

Return to the path at (1) and climb to the level above. The Josephine Randall Junior Museum (5) (run by the Department of Recreation and Parks) is open from 1000 to 1700 every day except Sunday and Monday and has a fine, small, mineral collection; an excellent little fossil hall; and two seismographs, one an ancient Bosch-Omori with all its working parts exposed (loaned from the U.C. Seismographic Station) and the other an operating standard recording drum. The east wing of the museum houses a live animal room with boa constrictors, raccoons, foxes, and a great variety of wild and domestic birds and small animals, many of which can be handled by youngsters.

The highest exposure of pillow basalt is at the west end of the museum parking lot (6). Its relations to the chert here are complicated. The high cliffs behind the museum expose more minor folds (7) whose short limbs bear the same relation to their hinges as do those on the cliff below. There are also many small faults, some with displacements of about 0.1 m, that die out completely within a few meters (8).

A paved path leads from the east side of the museum to a point on the east side of Corona Heights (9) that affords a magnificent view of San Francisco. From here you have an excellent overview of the structural blocks beneath the city (see Figure 5 and section above entitled Speculations on the Structure of the City). In the northeast corner of the city, largely hidden by the downtown towers, is the Nob Hill Block (V), mainly greywacke. Closer at hand, on the north side of Market Street, the New Mint Building sits on its grey-green hill of serpentine, part of Block IV, which extends southeast to in-

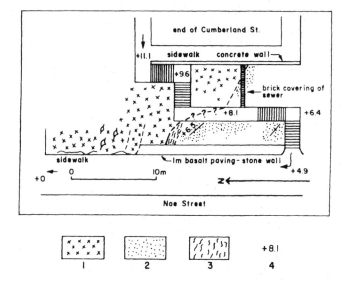

Fig. 11. Sketch map of geology at west end of Cumberland Street (locality C, Figure 8): **1**, greenstone (basalt);**2**, sandstone; **3**, sheared sandstone and shale; **4**, approximate height, in meters, of sidewalk above sidewalk on east side of Noe Street, at north end of outcrop.

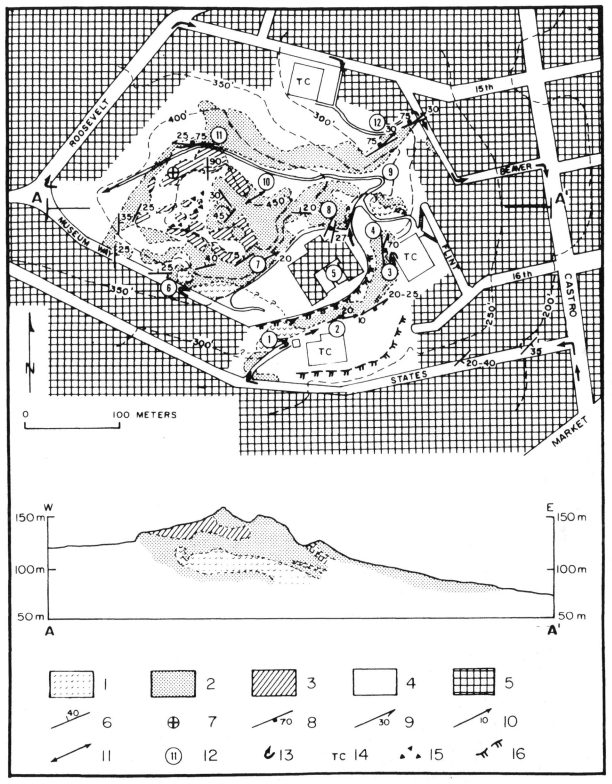

Fig. 12. Geology and points of interest in vicinity of Corona Heights Playground. Base map drawn from a California Department of Transportation (Caltrans) Bay Area Transporation Study (BATS) aerial photograph taken in 1965; 50-ft contours transferred by inspection from USGS topographic map of San Francisco north (7.5′) quadrangle; **1**, exposure of pillow basalt; **2**, exposure of chert; **3**, exposure of sandstone; **4**, covered; **5**, built-over areas; **6**, strike and dip of bedding; **7**, horizontal bedding; **8**, strike and dip of faults and joints; **9**, bearing and plunge of slickensides; **10**, bearing and plunge of hinge line of minor fold; **11**, horizontal hinge line; **12**, locality mentioned in text; 13, route of walking trip (shown only at critical corners); **14**, tennis court; **15**, large blocks of chert in colluvium; **16**, top of quarry wall (shown only south of Josephine Randall Junior Museum).

16

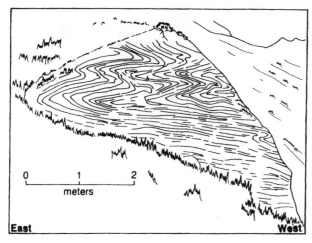

Fig. 13. Recumbent fold at locality 10, Figure 12, on north side of Corona Heights Playground, looking southwest along direction of horizontal hinge lines; sketched from a photograph.

clude Potrero Hill, the large hill with a gas tank peering over it, and Hunter's Point, the distant hill to the south of the large black gas tank. To the right of these hills is another line of hills, including Bernal Heights, Mission Heights, and Corona Heights—on which you stand. These, and the mountains behind them (Twin Peaks, Mt. Davidson, and Diamond Heights), are in the Twin Peaks Block (III), which is mainly chert and basalt. In the far distance to the south is the even crest of San Bruno Mountain, consisting of well-formed layers of turbidite sandstone and shale of Block I. The belt of melange beween Blocks I and III is hidden in the valley south of Diamond Heights.

The path turns sharply west atop a narrow wall between two quarries and ultimately enters a small high-level quarry that contains a pair of brick picnic fireplaces. Just west of the fireplaces, at (10), is an exposure of symmetric recumbent isoclinal folds in chert (Figure 13). West of this, where the path bends around the hill (11), sandstone is exposed along the path. This is part of a core of sandstone in what appears to be a tight recumbent fold (see cross section, Figure 12), for chert is exposed in the cliffs just below the path and can be traced beneath the sandstone around the west side of the hill, and the hilltop is capped by chert overlying the sandstone—chert that is continuous with the chert beneath the east half of the hill. Several dirt trails across the hilltop enable one to check out the field relations. It is tempting to relate this structure to the minor folds in the chert (for example, 10 is near the hinge of this structure), but the hinge line of the large structure (if it is a fold) appears to trend due north, whereas those of the minor folds in the chert trend northeast.

After exploring the hilltop, continue west across

the greensward and out the gate to the corner of Roosevelt and Museum Way. Turn right (north) on Roosevelt Way; right (east) on 15th; and about 3/4 of the block down 15th, just beyond the tennis courts, right again into Sidney Piexotto Playground. The high quarry wall on the southeast side of this playground (12) is an exposed fault surface that has some of the finest slickensides and mullions to be seen anywhere. These plunge about 30°NE. Rubbing the slickensided fault surface will enable you to determine that the northwest side moved relatively down and northeast. This is the best of numerous slickensided fault surfaces exposed in Corona Heights, most of them striking northeast. Most of the faults cannot be traced for any length. For example, this fault cannot be recognized in the quarry walls behind the junior museum on the other side of the hill, toward which it trends.

At the east end of the playground a path leads beside the fault to the Beaver Street. Turn right (south) on Beaver, past a fine road cut exposure of nearly flat-lying chert; follow Beaver to Castro; turn right (south) on Castro and walk back to Market. Take an eastbound Muni Metro streetcar to Church Street station, and walk one block north on Church to E, Figure 8, the last exposure on this trip. (To make the trip directly to this locality, take any westbound Muni Metro streetcar at Market and Van Ness and get off at either Duboce or Church Street station; if necessary, walk north on Church to Duboce.)

This is probably the most beautiful and informative outcrop of serpentine in San Francisco. The excavation has left resistant (and commonly rounded) blocks of massive serpentine projecting in relief from the matrix of intensely foliated slickentite (Figure 14, see section above entitled What the Rocks Look Like).

Bastite grains in the massive blocks appear to occur in crude layers, which may be inherited from an original metamorphic layering of the pyroxene grains in the ultramafic mantle rock that was hydrated to form the serpentine. The layering has different attitudes in different blocks, suggesting that the blocks rotated during the deformation that produced the slickentite.

Although foliation in the slickentite wraps around the massive serpentine blocks, a walk around this block and around the New Mint Building to the east should convince you that despite the local variability the regional foliation dips between 15° and 30°. (Note: The vacant land in front of this exposure is usually used for the sale of Christmas trees during December or may be used as a storage yard during reconstruction of the Municipal Railway. If so, the rocks can be examined close at hand on the east side of Church Street, on the west side of the block.)

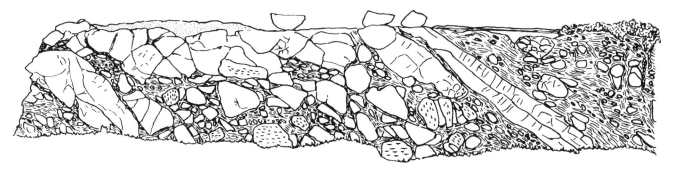

Fig. 14. View north to west half of serpentine exposure at Duboce and Church streets; sketched from photograph: fine lines, foliation in "slickentite"; blank, blocks of massive serpentine; short heavy lines in blank blocks, parallel alignment of "bastite" grains in massive serpentine. Exposure shown is approximately 4.5 m high and 24 m long.

This is the end of Trip 1. Take any streetcar back to Market and Van Ness and the Number 47 or 42 bus northbound on Van Ness to your origin.

Trip 2. To Fort Mason and Subducted Sandstone

This is a short trip to a pleasant picnic spot on the waterfront with fine views across the bay. Take the Number 47 trolleybus or Number 42 motor bus northbound on Van Ness to North Point (1, Figure 15) walk north on Van Ness to the start of the semicircular pier enclosing Aquatic Park Cove. The wooded hill on the left is Fort Mason (Black Point), embarkation point for soldiers to the Far East during World War II and now headquarters of the Golden Gate National Recreation Area (GGNRA), run by the National Park Service. Houses on the hilltop date from mid-19th century. On the right are Fontana Towers (apartments); Eastman Kodak Company; Aquatic Park, with bocce ball courts nearby; and a collection of ancient sailing ships and ferry boats (the San Francisco Maritime Park) across the cove. Swimmers can be seen in the cove, even in winter. The small circular building (2) just north of the railroad houses a snack bar and toilets, and beyond its is the Sea Scout boathouse.

At the entrance of the semicircular pier, which has spectacular views of San Francisco from its far end, walk a few steps to the left and to the end of a short pier (3) pointed directly toward Sausalito and Mt. Tamalpais. From here you can look back to the bold sandstone cliffs of Black Point, with sea caves and sea stacks. Northward is a view extending from the Golden Gate Bridge on the west to Alcatraz Island and the north end of the Berkeley Hills on the east (Figure 16).

The dark red mountains between the Golden Gate and Sausalito are the Southern Marin Block (see above). Chert generally comprises the crests

and west flanks of the ridges, and basalt underlies their east flanks. Valleys and passes are usually eroded along sandstone and shale. The south dipping thrust beneath this block passes through Sausalito.

Most of Marin Peninsula, visible to the right of the Southern Marin Block, is mapped as Franciscan melange and has the gently rolling, rounded topography characteristic of melange. Mt. Tamalpais, with its steep slopes and sharp crest trending northeast, is a topographic anomaly that has been interpreted by some to be a recent fault block, which I doubt. Salem Rice, of the California Division of Mines and Geology, discovered that the east peak is made of tourmalinized greywacke—probably the product of an ancient thermal spring—which gives it erosional resistance. The hills of Tiburon Peninsula, between Mt. Tamalpais and Angel Island, are a complicated melange with abundant exotic blocks of serpentine and glaucophane schist. This is the type of area for lawsonite (see Trip 6 for a discussion).

Angel Island consists of metamorphosed Franciscan containing such blueschist-facies minerals as aragonite, jadeite, and glaucophane. Its geology is described in Trip 6 and shown in Figure 37. Alcatraz Island consists of northeast-dipping, massive sandstone with thin shale interbeds, from which Valanginian (Early Cretaceous) fossils have been obtained [*Armstrong and Gallagher*, 1977].

From the foot of the pier, walk west up the paved road above the sea cliff. A few weathered outcrops of sandstone can be seen above the concrete beam retaining walls. Beyond the first bend (4) you can look down on sandstone in the sea cliff below. At the second bend (5) are spectacular views of the Golden Gate, with low hills of serpentine south of the Gate contrasting with the high cliffs of chert and pillow basalt to the north (Figure 16). Point Bonita, a classic pillow basalt locality [*Ransome*, 1893] can be seen through the gate. About 130 m beyond the second bend, take the steps (6) down to dock level.

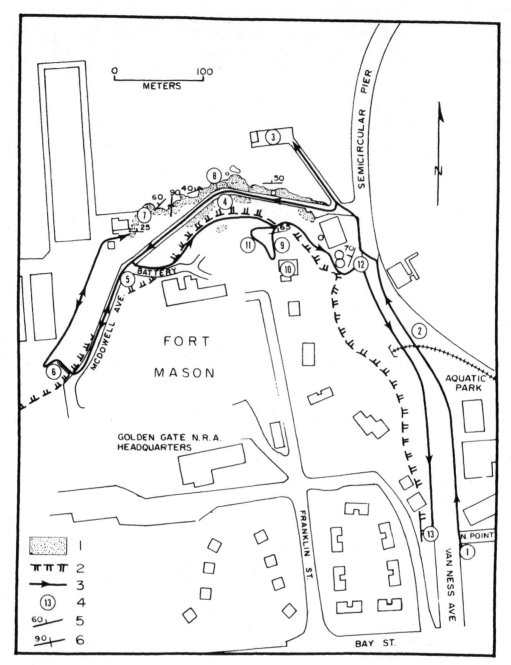

Fig. 15. Map of Fort Mason and vicinity, showing points of interest on Trip 2; drawn from Caltrans BATS aerial photograph taken in 1965: **1**, exposure of Franciscan sandstone; **2**, bluff top; **3**, route of walking trip; **4**, locality mentioned in text; **5**, strike and dip of bedding; **6**, strike of vertical bedding.

Dune sand plastered against the west side of Black Point is evident. Turn right at the foot of the steps and follow the base of the retaining wall to the water. Alcatraz Island is straight ahead. Go behind building 322 to exposures of sandstone (7).

At low tide you can scramble along the rocks almost to the northernmost point of the cliffs, but during most tides the only accessible sea cliff exposure is at the extreme west end, which you can reach by slipping between the railings along the walk. The sandstone is massive and thoroughly fractured on a variety of scales. Interlayers of siltstone and shale, although highly deformed and offset by small faults, generally indicate an approximately east-west strike and dip to the north. Indicators of which side is to the younger rocks are obscure and contradictory but seem to indicate younger rocks to the north (see Trip 3 for a discussion of these indicators). Among the minor structures visible are the small fold at (8) (see Figure 17) and complexly deformed shale interbeds (Figure 18). Subsequent to most of the deformation, veins of quartz (and possibly calcite) 1 mm-3 cm thick were deposited in fractures cutting the sandstone.

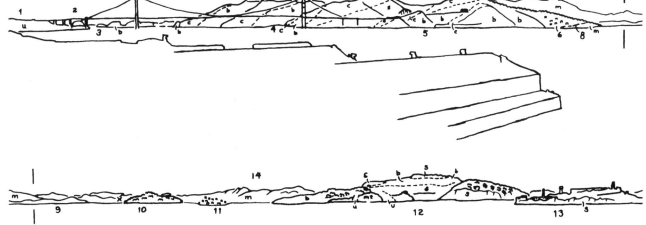

Fig. 16. The view from Fort Mason (locality 4, Figure 15); upper, west half; lower east half: **1,** Presidio; **2,** Fort Point; **3,** Point Bonita (behind Golden Gate Bridge); **4,** Kirby Cove; **5,** Fort Baker; **6,** Sausalito; **7,** Mt. Tamalpais (altitude 784 m); **8,** approximate trace of fault contact between Southern Marin Block and melange of Marin County; **9,** Richardson Bay; **10,** Belvedere; **11,** Tiburon; **12,** Angel Island; **13,** Alcatraz Island; **14,** Tiburon Peninsula; **b,** pillow basalt; **c,** chert; **m,** melange; **mt,** schist; **s,** sandstone; **u,** serpentine. (Sketched from photographs.)

Return by the same route to 5 (corner of Battery and McDougall), and take Battery to the right about 30 m, then take a concrete walk along the top of the bluff to the left. The walk passes a picnic area where the concrete path is blocked on the east by a chain link fence and steps go uphill and downhill. Take the uphill steps first, past a tiny outcrop of sandstone, to three important historical localities: the site of the house (9) occupied by John C. Fremont between 1859 and 1861; the Palmer House (10), built in 1855; and Bateria San Jose (11), a Spanish fortification established in 1797 to protect the harbor. Return to the cliffhead trail, take the steps downhill, and follow the concrete path around the east side of the hill. The retaining wall is built of slightly schistose sandstone, probably from Angel Island. Enjoy the fine views of Fisherman's Wharf, Aquatic Park, Ghirardelli Square, and Alcatraz and Yerba Buena islands. The parth descends by a flight of steps past round tanks to the north end of Van Ness Avenue. A tiny outcrop of weathered sandstone is at (12). The bus stop for the southbound Number 47 (Potrero) trolleybus, which will return you to your origin, is at (13).

Trip 3. Baker's Beach and Fort Point: A Trip to Melange and Serpentine

This trip is to some of the best and most easily accessible exposures of Franciscan melange and serpentine in California, at the northwest end of the diagonal serpentine belt across San Francisco. The bluffs between Baker's Beach and Fort Point (Figure 19) expose two bodies of serpentine (probably gently dipping sheets 100 m or more thick) with an intervening sheet of melange whose matrix is crushed shale and sandstone. The sheets and their foliation appear to strike generally to the north and to dip gently east, bending around at the north end to strike northeast and dip southeast. Much of the bluff is clothed in the debris of large landslides. The best exposures are in the headwalls of these landslides and along wave cut cliffs a few feet high along the shore. This trip is a beach walk to these exposures.

At the south end of the bluff the serpentine and melange are in fault contact with a thin, highly deformed body of red ribbon chert, which in turn is in fault contact with an excellent turbidite sequence that consists of about 80 m of predominantly sandstone beds that average a thickness of 2–3 m, strike northwest and dip about 45° northeast, and are overturned.

Like Trip 1, Trip 3 is divisible into several short segments. The entire trip involves a hike along 2.4 km of shingle and sand beach (some of the shingle consists of blocks the size of a house), across landslides, and up some short steep grades along narrow paths that may be muddy in wet weather. The maximum climb is about 85 m. Wear hiking or tennis shoes and clothes you don't mind getting wet or dirty, and carry a sweater or windbreaker as protection against stiff ocean breezes. The area is within the Golden Gate National Recreation Area (GGNRA), and geologic hammers should not be used, nor should specimens be collected from rock outcrops. The rocks and scenery are photogenic,

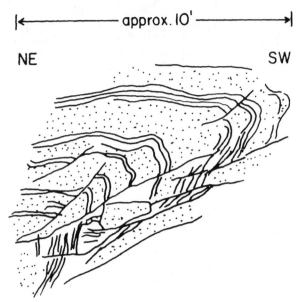

|← —— approx. 10' —— →|

NE   SW

Fig. 17. Field sketch of minor fold at locality 8, Figure 15, looking southeast.

and on a warm, sunny day there may be exposures other than those described in this guide.

The quality of exposures and ease of making the trip depend heavily on tides and seasons. Exposures are best during a lower low spring tide in the latter half of winter, when steep winter waves have lowered the beach as much as 2 m and exposed many critical relationships.

The footpaths to these exposures are all served by the same bus line—Muni Number 29 (Sunset). This line can be reached from Van Ness Avenue via the Number 38 (Geary), boarded westbound on the northeast corner of Geary and Van Ness (Number 38 starts at the Eastbay Terminal and goes up Market, past the Montgomery BART Station, to Geary). Transfer from Number 38 to Number 29 at the SE corner of Geary Boulevard and 25th Avenue.

There are four footpaths to the exposures: A, by way of Baker's Beach; B and C, down the central part of the bluffs; and D, underneath the Golden Gate Bridge at the north end of the bluff. For route A, get off the bus at Pershing Drive; for routes B, C, and D, get off the bus at the corner of Merchant Road and Lincoln Boulevard.

Each route between bus stop and beach is described together with outcrops and points of interest accessible along that route; then points of interest along the shore are described in order from south to north. Numbers in parentheses refer to localities on Figure 19. The turbidite sequence at the south end of the bluffs is accessible only via route A. Between these exposures and the melange and serpentine is a usually unavoidable 30-m climb through poison oak. Those susceptible to this pernicious plant should take routes C and D to exposures of serpentine and melange.

Routes to the Beach

Route A Cross Lincoln Boulevard, take any path downhill and through the picnic area in the Monterey pine (*P. radiata*) forest to the north end of the parking lot (restrooms and drinking water), and descend to the beach. Walk to cliff at north end of beach (easier and faster on the damp, packed sand left by the retreating tide). At (1) the seacliff consists of beds of massive sandstone 20 cm–4.6 m thick with 6-cm to 1-m interlayers of finely laminated to cross-bedded sandstone and shale. These are turbidity current deposits.

The criteria for recognizing turbidity current deposits were discovered by field observation and laboratory experiment in the period between 1950 and 1967 and were codified by A. H. Bouma into a progression of characteristic beds deposited by a single turbidity current during its passage and as it finally came to rest (cf. *Middleton et al.*, 1973, pp. 4, 29, and 57a). The Bouma divisions of the deposits of a single flow are given letter designations as follows (beginning at the base): A, massive graded sandstone, which may have pebbles or granules at the base; B, plane parallel laminae in sandstone; C, ripples, wavy or convoluted laminae in fine sandstone and siltstone; D, parallel laminae in siltstone; E, interturbidite deposits, generally shale. Unit A is the massive turbidity current itself, come to rest and settled on the seafloor. Units B, C, and D are the uppermost layers of the deposit, winnowed and redeposited by the turbulent eddies generated by the current. Unit E is the fine sediment that settled to the seafloor during intervals between turbidity currents. The turbidite beds at this locality exhibit most of the Bouma criteria, usually with Bouma units ACD or ACDE, and probably accumulated on the middle

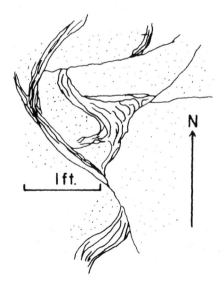

⊢ 1 ft. ⊣   N ↑

Fig. 18. Field sketch of deformed sandstone and shale at locality 7, Figure 15, looking vertically downward.

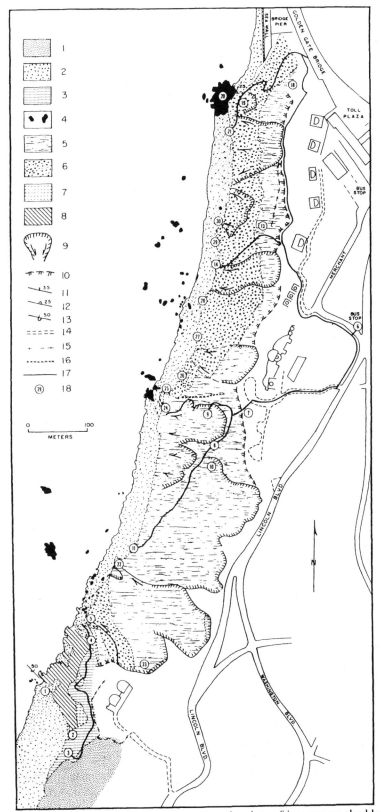

Fig. 19. Map showing geology of, access routes to, and points of interest on the bluff between Fort Point and Baker's Beach, Golden Gate National Recreation Area, destination of Trip 3; sketched from a 1965 Caltran BATS aerial photograph: **1,** Holocene dune sand; **2,** modern beach sand (shoreline is at lower low water); **3,** Pleistocene deposits (Colma Formation); **4,** boulders and rock outcrops on the beach and offshore; **5,** serpentine; **6,** Franciscan melange; **7,** chert; **8,** sandstone (turbidite sequence); **9,** active landslide; **10,** bluff top; **11,** strike and dip of bedding; **12,** strike and dip of foliation; **13,** overturned bedding; **14,** graveled roadway; **15,** dismantled fence line; **16,** crest of spur; **17,** access route to beach; **18,** locality mentioned in text.

22

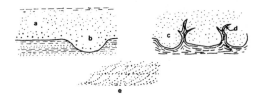

Fig. 20. Some criteria for telling if sedimentary rocks are right side up or overturned: **a,** graded sandstone bed with sharp contact with older underlying shale and gradational contact with overlying shale; **b,** filled channel; **c,** load cast; **d,** flame structure; **e,** cross bedding.

and lower parts of the submarine fan built by the turbidity current or near the mouth of the trench that these currents usually carve across the head of the fan.

The use of pillows in determining the direction to younger rocks, and hence in telling whether or not the rocks have been overturned, was given in the section above on The Origin of the Ocean Floor. Sedimentary rocks have features that make this determination possible and show that the beds at this locality are actually overturned. Some of these features are enumerated here and illustrated in Figure 20. When a slurry, such as a turbidity current, comes to rest, the larger and heavier grains sink to the bottom (unless they are less dense than the slurry), and the resulting massive bed is coarse at its base and grades upward to fine sand or silt at the top (a, Figure 20). Furthermore, it rests on underlying fine sediment. The current that deposited the coarse sediment may have first cut channels into the underlying sediment, truncating bedding and filling the channels back with the new deposit (b, Figure 20; this is the most reliable of these indicators). If the underlying sediment is still soft, the weight of overlying heavier sand may cause that sand to sink into the underlying sediment that displaces it. The bulbous projections of sand downward into the underlying fine sediment are called load casts, and the wisp-like projections of the displaced clay up into the sand are called flame structures (c and d, Figure 20).

Finally, sand transported along the bottom of a current of water or air commonly moves as ripples or dunes. These travel downcurrent as the sand grains eroded from their upcurrent slopes tumble down their downcurrent slopes. In cross section, a deposit formed of such ripples or dunes will consist of many sets of thin, inclined beds that curve to be tangent to the surface on which the ripple or dune was traveling and which are truncated by the base of material deposited on top of them. This inclined fine bedding, called cross bedding, tells not only whether the beds were overturned but also which way the current flowed that deposited them.

The best evidence here of overturning is over-

turned channels filled with coarse sandstone that truncates bedding in medium to fine sandstone, which is exposed only at low tide in late winter at the extreme southwest corner of the outcrop. Overturned flame structures and load casts are present on the first bedrock spur. Cross beds are equivocal, because of subsequent deformation, and large pebbles of coalified plant remains and shale appear to have floated to the top of the sand slurries. Fine coalified plant remains, abundant in the finely laminated sequences, perhaps indicate nearby land.

The beds are cut by numerous faults with displacement from a fraction of a meter to several meters. Veins of coarsely crystalline calcite fill many faults and other fractures.

Note the change in color, degree of consolidation, and spacing of fractures from unweathered sandstone at the cliff base facing the ocean to weathered sandstone at the cliff top, and also at beach level 30 m east along the back of the beach. Locality (2) shows that this weathering occurred long ago, perhaps in the mid-Pleistocene, several hundred thousand years ago, and certainly before the last (Wisconsinan) glaciation. Photography buffs may wish to record the delicate fretwork etched by salt spray in the variably cemented sandstone.

Immediately to the southeast is a complex exposure of Quaternary deposits (2) with talus and a thick clay-rich soil at the base, flanking contemporaneous lagoonal sands that are buried in turn by orange dune sands on which two very thick, clay-rich soil layers have developed. The thick soils dip about 20° south, away from the flank of the buried bedrock hill and parallel to the original pre-Wisconsin hillside. They are covered by unconsolidated grey Holocene dune sand. This exposure is evidence that the ocean has never stood higher on the northern tip of San Francisco Peninsula than it does now, else these deposits would have been removed by erosion. Implications of this exposure and others like it are discussed in Trip 5b.

The exposures of melange and serpentine must be reached by climbing over the top of the sandstone cliff, starting 45 m south of the Franciscan–Pleistocene contact, where a path follows a ravine (3) cut into the prominent soil zones. As you start up this path, look to the cliff to the north to see outcroppings of chert above and to the east of the sandstone. The buried mid-Pleistocene talus came from this outcrop.

Where the path reaches loose Holocene dune sand above the cliff-forming soils, a steep sandy path branches north to go, first, through lush poison oak, then through willows, yellow-flowering lupines, sea fig, coastal sage, and other sweet-smelling plants, then it levels off at the crest of a narrow spur carved along the highest exposures of the soils;

it finally bends north along an abandoned fence. Near the top, watch carefully for poison oak and rusty barbed wire. It is almost impossible to avoid poison oak here and on the plateau beyond.

The serpentine boulders at the top are not an uplifted cobble beach but part of the camouflage of World War II shore batteries. On the point overlooking the beach is a concrete pillbox, and below that is a concrete platform from which the sandstone ledges can be seen below. From the north side of this platform a path winds along the cliff face toward the beaches visible to the north. The path is mainly in buff-to-orange semiconsolidated dune sand, the continuation of the Pleistocene deposits that are on the south side of the hill. Chert crops out through the Pleistocene cover; its contact with sandstone is a steep fault parallel to the shore that truncates bedding in both sandstone and chert. Chert, together with large blocks of yellow jasper, is exposed at the narrow point at the north end of the cliff (4), where the path turns inland and drops precipitously to a little landslide valley. The high greenish-grey cliffs to the east are of serpentine, exposed in the headwall of a large landslide, and have orange Pleistocene dune sand at top. The exposures at the same level as the path, across the south branch of the little valley, are of melange interlayered with foliated serpentine.

The path follows the south edge of the landslide. Fresh cracks and low scarps may be present, especially after a rain. Avoid the landslide surface, especially during the rainy season, as it is soft and muddy, sometimes dangerously so.

Where Route A reaches the toe of the slide, highly deformed and boudinaged, thin-bedded chert and shale are exposed along a short path cut south into the cliff (5). From this path you can peer into the head of a deep narrow cove, accessible at low tide, where the complicated nearly vertical fault zone between the chert and sandstone is exposed for a height of 15 m.

Route B   From the corner of Merchant Road and Lincoln Boulevard (6), walk south about 50 m along the west side of Lincoln Boulevard and turn right on a paved road, keeping to the left of the pavement. The pavement ends at a graveled parking lot overlooking the beach (7). At the southwest corner of this parking lot a path leads through vegetation and down the bluff to the north end of a landslide headwall. The path to the beach skirts the top and south side of the landslide, where at (8) there is a view of the excellent serpentine exposures in the landslide headwall and where predominant foliation and flattening of the boulders apparently dips about 20° east. At (9) on the north side of the landslide, serpentine with nearly flat foliation may be studied near at hand. Shortly beyond (8), the

path joins another descending the cliff, then the spur and path turn abruptly south, with the basin of an ancient landslide on the east. The south-facing headwall of this basin (10) is a fine exposure of serpentine with huge, massive blocks containing bastite pseudomorphs. Predominant foliation dips east, as do variously colored bands of serpentine. The route descends southward, skirting the cliff top, to reach the south end of the beach at (11).

Route C   From (6), walk northwest (carefully) along the southwest side of Merchant. Where Merchant turns abruptly right toward the bridge toll plaza, go straight ahead to concrete shore batteries, and turn right along a red graveled path alongside the batteries. The Coastal Trail (well marked) branches off the path. Follow it past coyote bush, fennel, broom, and morning glory, down wooden steps, through a cypress grove, to a triangular flat at the top of a spur, where serpentine is exposed. The route to the beach is down this spur, first trending west, then bending abruptly south at (13) to angle down the bluff to the beach below. Just below (13) the highest outcrop of dark grey melange are exposed in the trail, but most melange is covered with landslide debris from the serpentine. The trail bends seaward just above the beach, is steep and slippery, and reaches the beach at a litter of cement blocks and rusty pipes (14).

Route D   The easiest route to the beach is to continue north along the Coastal Path on the cliff top. At one point the path is at the head of a landslide, where the torn end of a fence dangles over the cliff top. The trail joins a bicycle path, which you should follow downhill to bench just west of Golden Gate Bridge approach. From this bench, Route D goes directly downslope to a concrete pillbox (18), where it abruptly bends left along an abandoned roadway that switchbacks down the slope. The route here is mainly on melange. Serpentine is exposed in the bluff to the left. The path continues downhill through coastal scrub dominated by succulents and flowering plants and has magnificent views of the Golden Gate Bridge and Marin Headlands to the north. Serpentine in the path is mainly landslide debris. Heavy rains in January 1982 have reactivated a large landslide here, and the future route of the trail is uncertain. At present it is best to go from (19) directly toward Fort Point Rock (20) and find a well-trodden path to the shingle beach behind the rock. Proceed either north or south along the beach from there; this is most easily done at low tide.

Outcrops on the Beach

Between (5) and (22) is the toe of a large landslide. The base of the low bluff is accessible at low and intermediate tides and exposes landslide debris: an assemblage of rounded serpentine boulders in a matrix of extremely "fat," sticky, greenish to white clay, probably remolded from slickentite serpentine. The bright orange sand and clay just south of (22) are fragments of Pleistocene deposits, and soils slid from their original positions near (23) at the head-wall of the landslide, where similar materials are still in place.

At (22) a large tectonic block of well-cemented but highly deformed dark-grey sandstone and shale forms a 5-m-high promontory that juts into the sea. One can scramble around this promontory on the beach at low tide but must climb over the spur behind it at high tide. Foliated serpentine rests with irregular east-dipping tectonic contact on the sandstone block; the contact is exposed on the north side of the promontory. The sandstone block is in the melange, and a large block of basalt, chert, and pink limestone at water level to the north is in fault contact with the sandstone block along a 1-m-thick shear zone with masses of sandstone, shale, chert, and greenstone. Note: North of (22), landslides generated by the storm of January 2–3, 1982, have left mud temporarily covering some of the exposures along the beach described here.

Between (22) and (24) is a sandy beach about 275 m long and 30 m wide at low tide, backed by serpentine bluffs consisting of rounded blocks of massive serpentine embedded in pale green, intensely foliated slickentite. The foliation of the slickentite wraps around the boulders of massive serpentine but, in general, strikes northerly and dips gently east.

The beach boulders are massive serpentine polished by wave action. The bastite pseudomorphs appear, at first glance, to be mica. The boulders commonly have a delicate network of thin asbestos veinlets. Other boulders are yellow and red jasper, chert, greenstone, and intensely deformed sandstone and shale cemented by numerous quartz(?) and calcite veins.

Between (24) and (25), a steep cliff with huge blocks at its base and offshore separates the serpentine-backed beach from the melange-backed beach to the north. At low tide one can scramble around the blocks, but at high tide one must climb behind the sandstone block at (24), along a path that closely follows the serpentine-melange contact. For about 15 m north of this block the contact and foliation in the melange strike roughly north and dip 15°–40° east. At (25), contact and foliation bend sharply to strike about N 60°W and dip 80°N, and the contact climbs the bluff for about 20 m then bends sharply northward to a nearly horizontal attitude. These bends are probably kink folds rather than faults.

Between (25) and (26), foliation in melange at

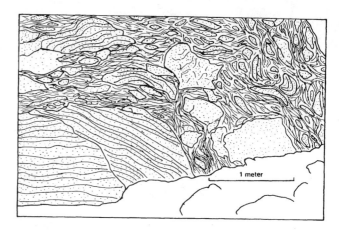

Fig. 21. Block of thin-bedded sandstone and shale with indistinct contacts with matrix; locality 26, Figure 19; sketched from a photograph.

beach level is nearly vertical, and melange consists of intensely sheared, fine, dark-grey sandstone and shale, apparently originally in thin beds. A zig-zag pattern in relict bedding and foliation suggests tight chevron folds that were broken along appressed hinge lines.

At (26), serpentine is interlayered with melange in layers 2–5 m thick. Foliation in the slickentite is diagonal to and truncated by contacts with melange, implying that foliation, and therefore serpentinization, predated the tectonic juxtaposition of these bodies. At the base of the bluff a body of nearly flat-lying but overturned thinly bedded sandstone and shale is nearly intact in its interior but crushed around its margins, so that its contacts with the enclosing melange matrix are indistinct (Figure 21).

At (27), blocks of thinly bedded sandstone and shale are embedded in a sheared melange matrix. The beds in one block appear folded into an anticline. However, cross bedding in a thin sandstone bed shows that the beds are overturned and that this is an overturned syncline. Shearing in the melange matrix between this block and a 5-m-high massive sandstone block 20 m north drapes over the blocks and is nearly horizontal between them.

At (28), a block of green metamorphosed gabbro about 20 m long and 5 m high above beach level is exposed at the cliff base. Melange matrix wraps around and is infolded or infaulted into the rounded upper surface of this block, and patches of the intensely deformed black matrix containing sheared off and deformed fragments of gabbro are preserved on this surface.

The huge sandstone block at (29) exposes a 6-m-thick turbidite sandstone with very coarse sand at its base, resting with locally faulted erosional inconformity on fine sandstone and remnant bodies of shale of the Bouma C, D, and E members, all right side up. Large blocks of green chert that had been tight-

Fig. 22. Golden Gate Bridge and Marin headlands from the west side of Fort Point. (Photograph by Wayne Monger.)

ly folded, broken, then recemented, are at the back of the beach at (30). The bluff between (20) and (30) is mainly landslide debris, principally from melange. North of (21), however, the landslide debris consists mainly of serpentine.

At (20), at Fort Point Rock, one can scramble north around the base of the bluff to a shallow cove whose far side is the granodiorite block seawall of the massive concrete south anchor for the bridge cables. Here melange rests on the serpentine that makes the foundation for the south tower and bridge anchor. The contact of melange on serpentine follows the shore northeast, then rises gradually toward the top of the concrete bridge anchor. Foliation in the serpentine commonly wraps around massive serpentine blocks but generally appears to strike N 30°–60°E and dip 15°–40° south.

From here one can see the north side of the Golden Gate, where the geology is completely different (Figure 22). In the mountains around Kirby Cove and the forested valley just west of the bridge, coherent bodies of pillow basalt and chert (Trip 5) strike southeasterly, directly toward Baker's Beach, and dip southwest. A major structural discontinuity beneath the Golden Gate is indicated.

Trip 4. A Sedentary Survey of the Structure of the City (With Side Trips Afoot)

This is a circle tour of San Francisco by Muni bus to see, mainly at a distance, outcrops of each of the five structural belts that underlie the city (Figure 5). This tour also explores the structure of the city in human terms, and as geologists read rock history from structure and decoration—such as fossils—so one can read the history of settlement from the

structure and decoration of houses and buildings. To enliven the intervals between outcrops, these features are summarized below.

The core of the tour involves four bus routes (Nos. 44, 52, 15, and 42) and starts along Sixth Avenue in the northwest quarter of the city (Figure 23), goes south through the central highlands to the southeast corner, north around the eastern and northern edge, and south on Van Ness Avenue. From Van Ness Avenue to the No. 44 line, many line are available, but only three are described here.

The outcrops whisk by so quickly that they can only be glimpsed from the bus. Hence four side trips are included at appropriate places for a closer look at the rocks and their structure.

History and Social Geography

Although the first settlements within what is now San Francisco were the Presidio and Mission Dolores, established by the Spanish in 1776, the germ of the present city was established in 1832 on Yerba Buena Cove just south of Telegraph Hill. When California was ceded by Mexico to the United States in 1846, this village had a few hundred inhabitants. Upon the discovery of gold in 1848, people poured into California from around the world, and San Francisco exploded from less than 1000 people in 1848 to 34,000 in 1852 and 150,000 by 1870. Its present population is about 700,000.

By 1853 the city extended about a mile west and south of its point of origin, and by 1870 it covered the area east of Van Ness Avenue. Streets were laid out half way across the peninsula and through the Mission district. By 1895, building had spread four miles west and south from the original settlement.

The earthquake and fire of 1906 destroyed most of the area east of Van Ness and north of Mission Bay, but the city was quickly rebuilt and by 1915 extended to the ocean on both sides of Golden Gate Park. A streetcar tunnel opened in 1917 beneath the Twin Peaks made access easy to the dune field west of Twin Peaks, and the field was covered with houses by 1960. The major changes after World War II have been (1) growth of continuous suburbs in the lowlands surrounding the Bay; (2) building of freeways; (3) leveling of buildings in a wide area between Franklin and Divisadero (centered on Geary) for redevelopment (1948–60); And (4) the growth of giant boxlike skyscrapers, beginning about 1955.

The financial, commercial, governmental heart of the city extends two to seven blocks north of Market and one to three blocks south, between Van Ness and the Bay: business and finance being concentrated east of Kearney, government west of Taylor, and shopping between (Figure 23). However, many ar-

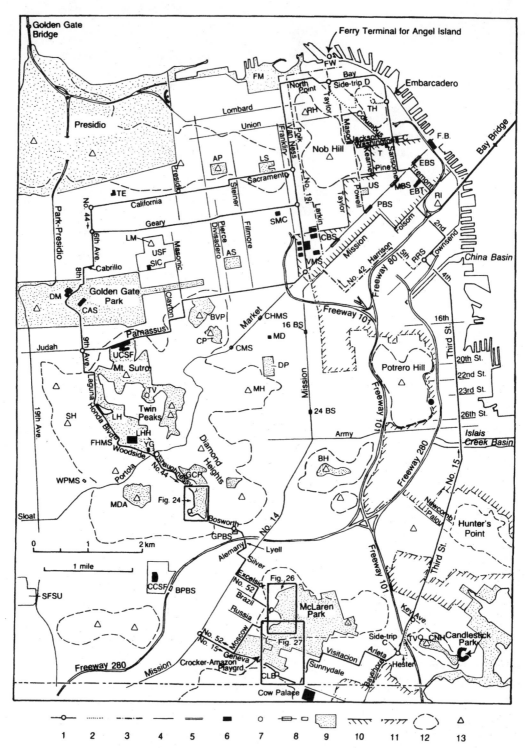

Fig. 23. Route of Trip 4, showing bus lines, transfer points, localities mentioned in text, and locations of Figures 24, 26, and 27. Also shown are bus and cable car routes to the ferry terminal for Angel Island (Trip 6): **1,** route of Trip 4, showing transfer point and stop for side trip; **2,** side Trips C and D; **3,** routes to Angel Island Ferry Terminal; **4,** other streets and continuations of bus lines; **5,** freeways; **6,** landmark buildings; **7,** TV towers; **8,** BART and Muni Metro stations; **9,** park or public open space; **10,** boundary of financial, commercial, and governmental core; **11,** boundary of industrial and warehouse district; **12,** boundary of upland area; **13,** hilltop or summit. **AP,** Alta Plaza; **AS,** Alamo Square; **BH,** Bernal Heights; **BPBS,** Balboa Park BART station; **BVP,** Buena Vista Park; **C,** Custom House; **CAS,** California Academy of Sciences; **CBS,** Civic Center BART station; **CCSF,** City College of San Francisco; **CHMS,** Church Street Muni station; **CMS,** Castro Street Muni station; **CNH,** Candlestick Hill; **CP,** Corona Heights Park; **DM,** De Young Museum; **DP,** Dolores Park; **EBS,** Embarcadero BART station; **EBT,** East Bay Terminal; **FB,** Ferry Building; **FHMS,** Forest Hills Muni station; **FM,** Fort Mason (Golden Gate National Recreation Area Headquarters); **FW,** Fisherman's

teries, such as Market, Mission, Van Ness, Polk, and Geary, are commercial throughout their length, and every neighborhood has its shopping area. The industrial area covers the flatlands bordering the bay that are south of the financial district and generally east of James Lick (101) Freeway (Figure 23). Multistory apartment buildings dominate the area south of Union Street, east of Divisadero, and north of the financial and other districts, and Market Street; they are also common elsewhere in the city. The remainder of the city not given to public parks is covered by single-family houses or by flats (two- to three-story houses with separate dwellings on each floor).

Affluent San Franciscans live along the north edge of the city, near the Golden Gate or the Presidio, and on the hilltop areas northwest of Mission Street (e.g., Telegraph, Russian, and Nob hills, Mission Heights, and St. Francis Woods). Those of middle income live mainly in the endless rows of houses on the ocean side of the city, either west of Twin Peaks (Sunset district) or north of Golden Gate Park (Richmond district), and in the hillier areas between Mission Street and the affluent enclaves of the Twin Peaks highlands. The southern and eastern parts of the city are sturdily working class, and the poor are crowded into such areas as the lowlands of the Western Addition around Divisadero and Fillmore streets south of California; Chinatown; the western part of the commerical area, where cheap hotels abound; the Mission District, and Hunters Point. The socioeconomic geography does not have sharp boundaries, and much more than in most American cities, rich and poor, black and white, straight and gay, live side by side in relative harmony.

On Domestic Architecture San Francisco is famous for the exuberance of its Victorian domestic architecture, built largely by artisans and contractors whose instructions came out of building-supply catalogues and plan books and whose inspiration was unimpeded by academic training in architecture or aesthetics. The balloon-frame construction, invented in 1839, enabled a few moderately skilled workmen to erect a study wooden house from precut lumber in a few days; in San Francisco, practically all private dwellings under five stories high are frame houses built of redwood and Douglas fir from forests of the nearby Coast Ranges. The low cost of

houses and lots enabled a large proportion of the artisan class to own their homes. The scroll saw, lathe, and steam press could cut or mold redwood into any desired shape, and prior to 1915, fantastically cut wooden pieces were turned out by the thousands to decorate the houses. Beginning about 1900, the taste for elaborate decoration waned, and by 1915, most houses had relatively smooth walls, many covered with stucco or asbestos sheeting, a style that has persisted to the present.

The standard lot in San Francisco is 25 feet wide and 75 to 100 feet deep, barely enough for the house, which was built right against its neighbors and, commonly, flush with the sidewalk. The yard, if any, was in back. The only visible wall on which to express taste and affluence was the front elevation (the wealthy, of course, could consolidate many lots to make room for detached villas and mansions).

Most of the 19th-century houses had gabled roofs with ridgepoles, but a flat-topped false front with a prominent, heavily decorated classic cornice was the fashion. Sometimes, a fake mansard roof fragment, called a "French cap," was added. Bay windows were also popular, with the sides of the bay either inclined or perpendicular to its front; bays had decorative entablatures or pediments across their tops. Inside, the lower floor had a hallway along one side, into which opened parlors and dining rooms separated by sliding doors; kitchen and laundry were at the back and bedrooms on the second floor. (Curiously, after flat roofs became standard, gabled false fronts and French caps became popular.)

The style of decoration makes it possible to date a house almost to the decade. The styles since 1870 are summarized below (the first three are the Victorian).

1. Italianate (1870–1884): Houses with false fronts—flat or with bays with inclined sides. Dentiles (square toothlike decorations) project downward from heavy classic cornices. Windows are rounded at the top, porches have Corinthian columns, and decorations are rectangular and projecting, especially on corners, in imitation of stone. The name comes from the supposed resemblance to Italian country villas.

2. Eastlake (now called stick style) (1880–1891): corniced false fronts with strong emphasis on vertical lines and on the structural character of wood.

Wharf; **GCP,** Glen Canyon Park; **GPBS,** Glen Park BART station; **LH,** Laguna Honda; **LHH,** Laguna Honda Home; **LM,** Lone Mountain; **LS,** Lafayette Square; **MBS,** Montgomery BART station; **MD,** Mission Dolores; **MDA,** Mt. Davidson; **MH,** Mission Heights; **PBS,** Powell BART station; **RH,** Russian Hill; **RI,** Rincon Hill; **RRS,** Caltrans (Southern Pacific) Railroad station; **SFSU,** San Francisco State University; **SH,** Sunset Heights; **SIC,** Saint Ignatius Church; **SMC,** Saint Mary's Cathedral; **T,** Transamerica Pyramid; **TE,** Temple Emanu-el; **TH,** Telegraph Hill; **UCSF,** University of California Medical Center; **US,** Union Square; **USF,** University of San Francisco; **VMS,** Van Ness Muni station; **WPMS,** West Portal Muni station; **YG,** Youth Guidance Center; **16BS,** 16th Street BART station; **24BS,** 24th Street BART station.

Rectangular windows, bays with perpendicular sides, pediments over windows squeezed into high little triangles. Extravagantly decorated with turned and scroll-cut wood and with plaster. Named for Sir William Eastlake, a prominent 19th-century English furniture designer, who deplored these decorative excesses in his name.

3. Queen Anne (1886–1901): Gabled fronts, strong emphasis on the horizontal and on the craftsman's ability to mold wood into any desired shape. Hence, rounded corner towers with conical turrets, arches over porches and windows, and decorative open screens of beaded dowels. Walls are commonly shingled. The style has nothing to do with Queen Anne, who reigned from 1702 to 1714.

4. Edwardian (1895–1915): Decoration of the plain clapboard surface supressed, except for entablatures with dentils and egg-and-dart moldings on each floor, in imitation of stone; bay windows with inclined sides; outside staircases in front of flats.

5. Craftsman or Western stick (1895–1915): A back-to-basics style with wooden framing prominently displayed, for example, heavily carved rafters to the edge of projecting eaves of a gabled roof. Note: since 1895 there has been much pretentious and slavish antiquarianism, with fake Tudor, Palladian, and Renaissance chateaux springing up in the enclaves of the wealthy.

6. Mission, Spanish colonial, and Mediterranean (1910–1935): Flat, plain stuccoed walls, in imitation of stone or adobe. Parapet may be decorated with a sloping red-tiled roof fragment like a French cap, only flatter, but has no cornice; arched doorways.

7. Bauhaus box (1930 to present): Rectangular boxes, with or without overhanging eaves and with strong horizontal lines. Windows are flush with the walls, may be horizontal ribbons, and may wrap around corners. Undecorated walls are painted, stuccoed, or shingled, for esthetics are conveyed by shape, not by decoration. For customers with fancier tastes the contractor adds a French cap and a nonfunctional balcony with ornamental iron railing in front of an arched picture window. Such houses are stuccoed or painted in light or pastel shades. The vast monotonous tracts of these row houses in the hills southwest of the city were the inspiration for Malvina Reynolds' song "Little Boxes." Recently plywood has replaced the now scarce redwood clapboards, and wood may be left unpainted.

Although one style usually dominates each neighborhood, the duration of its build-out may span several decades, so styles are intermingled. Furthermore, Victorian decoration fell out of favor from 1920 to 1960. I recall thinking in the 1930's that Victorian houses were bizarrely ugly. So did many of their inhabitants, who tired of the frequent re-painting. The decorations were stripped, and walls covered with stucco or asbestos siding. Hence many houses are Victorians in disguise.

Garages became mandatory about 1920 and occupy the ground floor beneath the living quarters. Older houses rarely have them.

Routes from Van Ness Avenue to the No. 44 Line

Ten transit lines connect Van Ness with the No. 44 line before the first outcrop on the latter. For brevity, only three are described here: No. 1 (California) for architecture, No. 38 (Geary) for convenience, and No. 6 (Parnassus) for the view. The others are shown on the Muni route map, and houses on them are described in architectural guides.

Landmark buildings visible from all three routes are Saint Mary's Catholic Cathedral (SMC, Fig. 23), built in 1971 and composed of four marble hyperbolic parabaloids 190 feet high that cast interesting shadows on each other; the red-tiled orange Byzantine dome of Temple Emanu-el (TE), built in 1924; St. Ignatius Church (SIC), of 17th-century Renaissance design, built in 1914 with Campanile, twin towers and a golden dome; the University of California Medical Center (UCSF), a cluster of modern hospital towers at the north base of Mt. Sutro; and Lone Mountain (LM), crowned by buildings of the University of San Francisco, the site of an early report on the mechanics of landslides [Cogan, 1936].

Catch the No. 1 line on the north side of Sacramento, east of Van Ness. Westward on Sacramento, it goes through a well-to-do district on the south side of Pacific Heights, passing Lafayette Park. Near Alta Plaza it jogs one block left on Steiner to California and stays on California past 6th Avenue. Between Franklin and Presidio, interesting Victorian houses of all three styles are interspersed with apartment buildings erected after 1920. West of Presidio, San Francisco's cemeteries of the 19th century were on the south side of California; the dead were exiled from the city in 1914, but removal was not completed until the late 1940's. Hence this area of Bauhaus and modified Bauhaus boxes. Beyond Arguello, turn-of-the-century architecture predominates. The transfer point to the No. 44 bus at California and 6th is in an undistinguished neighborhood shopping area, but a much more exciting shopping area is one block south along Clement. Geologically, the route is over sandstone and shale of the Nob Hill Block east of Presidio and presumably over the serpentine belt to the west, but all bedrock is hidden beneath dune sand and solidly covered with houses, streets, and gardens.

Board the No. 38 bus on the north side of Geary

just east of Van Ness. It proceeds along Geary to 6th. This is the main east-west artery of the north half of the city and passes through the center of 1950's to 1970's redevelopment. There are no Victorian relicts. Saint Mary's Cathedral is on the left at the top of the first hill, and beyond, to Fillmore, are garden apartments. The Japanese Cultural and Trade Center is on the right. On the southwest corner of Geary and Fillmore is (or was) Fillmore Auditorium, site of fabled rock music concerts in the 1960's. Beyond Divisadero, Geary climbs past Kaiser Hospital to the hilltop, beyond which is a vista of unrelieved commercialism to 6th. The rocks crossed are the same as those beneath California Street, with the same wealth of exposures.

Board the No. 6 bus on the north side of Market West of Van Ness, beside the Metro entrance. No. 6 turns right onto Haight, which has a rich variety of well-preserved and colorful Victorians, mainly stick and Italianate east of Pierce and Queen Anne to the west. Beyond Baker, Buena Vista Park is to the south. West of the park is the Haight-Ashbury district, which was a benignly integrated neighborhood in the 1950's and for that reason attracted the flower children of the 1960's, who made it famous with their be-in's, love-in's, etc., in Golden Gate Park nearby. The flower children attracted drug pushers and other predators in the 1970's, and the vibes turned bad. The district is now being revitalized as a healthy and interesting shopping area.

No. 6 turns left on Masonic, past some of the most interesting Queen Anne houses in the city (on the left), jogs on Frederick and Clayton to Parnassus, and at the east end of the University of California Medical Center it passes a chert outcrop (on the left). North (right) of the medical center are spectacular views of the Golden Gate Bridge and the Marin County Mountains. From here, San Francisco looks like a forest because the trees of Golden Gate Park block out the buildings. Beyond the Med Center, Parnassus becomes Judah. Transfer to the No. 44 bus at Ninth Avenue.

The No. 44 (O'Shaughnessy) Line (Figure 23)

The No. 44 line starts at a traffic island in the middle of California Street just west of 6th Avenue, stops at the corner of Geary (transfer from No. 38), jogs west of Cabrillo to 8th and south on 8th to Golden Gate Park. South of Geary, many houses built between 1895 and 1915 have sturdy, Craftsman-style rafters beneath overhanging eaves. Lone Mountain (LM) is to the east.

The route through Golden Gate Park is past the concourse with the De Young and Asian Art museums and Japanese Tea Garden on the right and the California Academy of Sciences (aquarium, plane-

tarium, geologic, biologic, and anthropologic exhibits) across the concourse on the left. Beyond the concourse is the Strybing Arboretum. All these wonders are a temptation to leave the bus here, but they are not part of this sedentary survey.

No. 44 leaves Golden Gate Park on 9th Avenue (transfer from No. 6 at Judah) with views of Mt. Sutro, covered by Sutro Forest (planted eucalyptus) and crowned by a 900 foot TV tower. The U. C. Medical Center is at Sutro's north base. No. 44 turns east on Lawton, and exposures of chert and weathered basalt at the base of the mountain can be glimpsed as the bus turns south onto Laguna Honda Blvd. The steep sandy hillside to the west is the slip face of the Holocene dune field, which covers most of San Francisco from here to the Pacific. The slopes of Mt. Sutro on the east, across Laguna Honda reservoir, expose chert overlying sandstone, probably on a thrust fault. The reservoir was a natural lake.

Beyond the reservoir are the grounds of Laguna Honda Home, San Francisco's hospital for the elderly. The grounds are on a brownish mid-Pleistocene aeolian deposit, the Colma Formation. Just past Forest Hill Station (on the right—connection to Muni Metro) the bus makes a double bend to the left to climb up Woodside Avenue to a pass nearly 200 m (600 ft) above sea level that is surrounded by steep hills rising 200–300 feet higher. The hills are commonly crowned with chert, and their sides are chert or greenstone, whereas the passes are underlain by shear zones. Near the top of the hill on the left is the euphemistically named Youth Guidance Center, behind which are exposures of sandstone and basalt.

At the pass the bus crosses Portola Boulevard, the main route from Market Street to the southwest part of the city, and starts downhill on O'Shaughnessy Boulevard along the west wall of Glen Canyon. If you plan to take side trip A, alert the driver that you wish to get off at Malta.

Scenery and geology pass with dizzying rapidity on the winding downhill run. To the north are Twin Peaks, whose summits and north sides are chert and south sides basalt. Across the canyon is Diamond Heights, crowned by condominium apartments in the brown style of the 1970's. To the west is Mt. Davidson, at 283 m (929 ft) the highest point in the city. The canyon below is Glen Canyon Park. On its far side, natural exposures of chert, making lines of crags and cliffs diagonally downward to the north, are practice grounds for rock climbing. Road cuts on the right expose chert in fantastic folds.

Side Trip A (Figures 24 and 25) Debus at Malta, cross to the east side of O'Shaughnessy, and walk north to study at leisure the outcrops you just flew past. The description is keyed to the light poles on the west side of the street; these are labeled in Figure 24. Light pole 1 is at the corner of Malta and O'Shaughnessy. Light pole 2 is at the mouth of a steep draw partly filled with stony colluvium. Light pole 3 is in front of a flat bench, behind which the cliff exposes intricately folded, thin-bedded, red chert. Fold hinges here are nearly vertical.

Behind light pole 5 the thin-bedded chert appears to grade northward into thicker-bedded yellowish-brown chert, although the contact is covered and may be a fault. Massive white- to yellow-weathering chert makes craggy outcrops on the point above. Just north of light pole 5 the chert is thrown into bizarre upright chevron folds with amplitudes and wavelengths of 2–3 m. Hinges plunge 35°–60° N45°–75°W. A broad swale lies between light poles 5 and 6, but the exposure is essentially continuous.

The most spectacular folds are along the base of the cut for 150 feet (50 m) south of light pole 6 (see Figure 25). Their hinges plunge 30° N60°W. The south limbs of anticlines are shorter than the north limbs; nevertheless, the average trace of bedding in the cliff is nearly horizontal.

Between light poles 7 and 10 the massive cliff-forming chert descends to road level and appears to be in steep fault contact with thinner-bedded more reddish chert on the north that dips 20°–30° northwest. The massive chert shows bedding if examined closely but seems to lack shale. The thin-bedded chert is overlain in turn by more massive chert that descends beneath road level at light pole 11. The fault contact between this chert and sandstone is exposed in the next road cut north and dips gently south beneath the chert.

This large outcrop of chert is part of a northward-dipping slab that ends abruptly against sheared sandstone at the top of the craggy outcrop across the canyon. Similar slabs of chert crown Diamond Heights, Mt. Davidson, the Twin Peaks, and Mt. Sutro. Some of the basalt on which the chert was originally deposited is locally present on the south sides of the slabs, for example, one-half block up Malta Avenue. The minor folds, probably kink folds originally, are extremely disharmonic, changing character or dying out altogether within a few meters across the bedding. This together with their marked asymmetry suggests that they are drag folds rather than compressional phenomena.

To study the folds close at hand, quickly cross O'Shaughnessy between light poles 5 and 6, watching carefully for traffic. When you have satisfied your curiosity, return to the bus stop and take the next bus south to Glen Park station.

The No. 52 (Excelsior) Line (Figure 23)

At the bus stop beside Glen Park BART station, on the south side of Bosworth, you transfer to a No. 52 bus southbound. While waiting for the bus, examine the granite cylinder in the station plaza that is inscribed with its exact latitude and longitude (37°44'08"N, 122°26'00"W). The plaza has a distant view of Diamond Heights and the chert outcrop you just visited.

Beyond the Freeway, the bus is on Lyell, turns right on Alemany, left on Silver, right on Mission, and left again on Excelsior. The route on Mission is through one-story Mission-style cottages of the 1920's and 1930's. The mural on the left as the bus turns onto Excelsior is a neighborhood art project. A few blocks farther, No. 52 jogs one block to the right on Maple to Brazil and turns right again from Brazil onto Prague. Side trip B begins at the corner of Brazil and Prague.

Most houses in the Excelsior district are single-family cottages, mainly in Mission style, built in the 1920's and 1930's. A few, lacking garages and of Victorian or Edwardian appearance, were built by pioneers in this neighborhod, some as early as the late 1880's.

Those not wishing to take side trip B should stay on the No. 52 bus to the corner of Mission and Geneva and should skip to the end of the description of this side trip.

Side Trip B (Figures 26, 27, and 28) This side trip begins in the northwest corner of McLaren Park, where blocks of serpentine, schist, gabbro, basalt, chert, and sandstone are embedded in a melange matrix of sheared shale and sandstone, and ends at Castle Lanes Bowling Alley on Geneva Avenue in an abandoned quarry whose walls expose turbidite beds of the San Bruno Mountain Block, which according to some is an outlier of the Great Valley Sequence (see Trip 7). The walk is 2.7 km (1.7 mi) long and involves climbs of no more than 46 m (150 ft). Neither it nor side trip C should be made alone.

From the bus stop (Locality 1, Figure 26), walk north on Prague. At the end of the street (2) is fine-grained phyllite or semischist with minor folds. Take the steps and path northeast into the park. Water oozing from the cut and reeds beside the path suggest that the outcrop of schist is embedded in impermeable crushed shale. In about 30 m (100 ft) the path is joined by another from the west and branches to the east. Take the northeast (left) branch over the hilltop. At (3) this crosses a path leading from John K. Shelley Drive to a water tower (4) on the hilltop.

From the base of this tower, much of San Francisco can be seen. To the northwest are Mt. Davidson, Mt. Sutro, the Twin Peaks, and Diamond Heights. To the north is Bernal Heights, its summit a flat slab of red chert. Behind Bernal Heights on the left are the towers of downtown San Francisco and on the right the grey-green hulk of Potrero Hill, main-

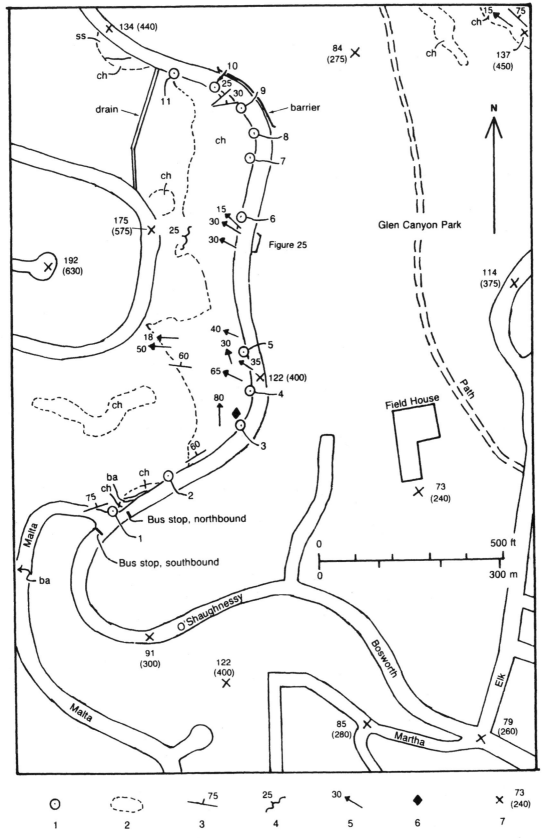

Fig. 24. Geologic sketch map of part of O'Shaughnessy Boulevard and Glen Canyon Park, showing route of side Trip A: **1,** light pole mentioned in text; **2,** boundary of outcrop; **3,** strike and dip of bedding in chert; **4,** average strike and dip of irregular bedding; **5,** bearing and plunge of hinge line of fold; **6,** location of vertical hinge line; **7,** spot elevation in meters (feet), estimated from topographic map; **ch,** chert; **ba,** basalt; **ss,** sandstone.

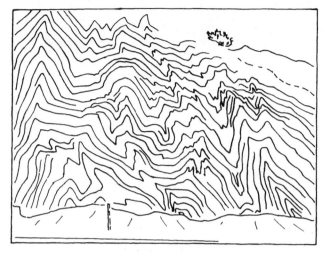

Fig. 25. Folds in chert on O'Shaughnessy Boulevard, looking west. Dimensions of area of cliff face shown, approximately 14 m wide and 8 m high. Only every third or fourth chert bed is shown. Traced from a photograph. For location see Figure 24.

ly serpentine. Farther right is Islais Creek Basin and Industrial Park, and to the right of that, almost hidden behind cypress trees, is Hunters Point—the southeast end of the serpentine belt. To the east are the hills of McLaren Park, in the melange belt, and Candlestick Hill, high and red, the southeast end of the Twin Peaks Block. San Bruno Mountain is the high mountain to the south, and the hills in front of it, decorated by crazy lines of "little boxes," are also in the San Bruno Mountain Block. The large building crowning the knoll on the skyline to the west is San Francisco City College (SFCC).

The hillside west of the tower has been cut away to make a flat for the houses below. Most of the rock in the cut (invisible from here) is sandstone with a strong schistosity, characteristic of textural zone II (see Trip 6), but at its north end are exposures of amphibolite, basalt, and serpentine (5), reached by continuing north along the path by which you entered the park.

North of (3), Shelley Drive curves east around a tree-covered knoll. Road cuts into this knoll expose schist and schistose sandstone. The best exposure is at (6).

Across Shelley Drive from (3) the path leads down by a series of broad steps to the shores of an artificial pond. Boulders in a pile (7), presumably excavated from the pond, are of basalt, serpentine, and schist. But the squared blocks of granite and diorite paving the pond spillway are surplus curbstones quarried in the foothills of the Sierra Nevada. The path around the west side of the pond passes blocks of basalt, diabase, and schist, some still embedded in the melange, as well as areas where the dark-grey clayey matrix oozes from the bank (8).

Return to (3), and take the pathway on the hillside east of Prague to the corner of Brazil and La Grande (9). Various boulders uncovered from the melange decorate the steps at the corner. Walk one block south, past Luther Burbank School, then east (left) on Persia for about 30 m (100 ft), cross Persia on the crosswalk at (10; Figure 26 and 27), and go south along the east side of Sunndyale. Blocks exposed north of Persia at the crosswalk are basalt and sandstone.

As Sunnydale Avenue bends southeast around the hill, its road cut is a fine exposure of melange. The intense shearing of the dark grey shale and sandstone matrix strikes roughly N 40°W and dips about 65° northeast. Embedded in this sheared matrix are large and small lozenge-shaped boulders of basalt and sandstone. At the south end of the first road

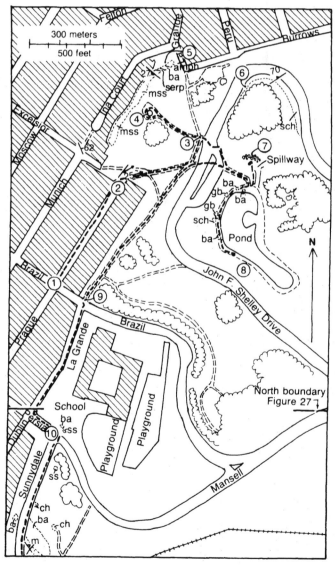

Fig. 26. Map showing route and points of interest on the north part of side Trip B. For explanation see Figure 27. (Sketched from a 1965 Caltrans BATS aerial photograph.)

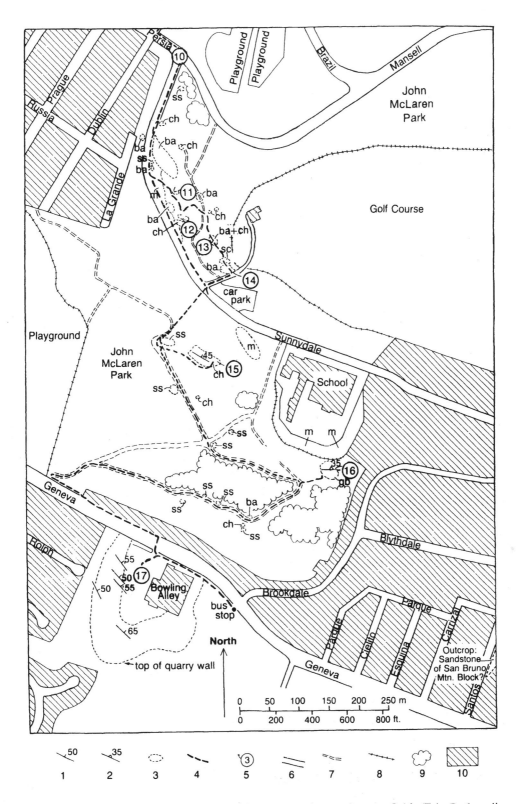

Fig. 27. Map showing route and points of interest on the south part of side Trip B: **1,** strike and dip of bedding in sedimentary rocks; **2,** strike and dip of schistosity of foliation in schist, melange, and gneissic gabbro; **3,** limit of outcrop or of exotic block; **4,** route of side trip; **5,** point of interest or locality mentioned in text; **6,** street; **7,** path; **8,** fence, **9,** grove of trees (shown only along route of side trip); **10,** built-up areas (houses, buildings, schoolhouse, and private yards); **amph,** amphibolite; **ba,** basalt; **ch,** chert; **gb,** gabbro; **m,** melange matrix; **mss,** meta-sandstone; **sc,** silica-carbonate rock; **sch,** schist; **ss,** sandstone. (Sketched from a 1965 Caltrans BATS aerial photograph.)

Fig. 28. Turbidite beds exposed on west wall of abandoned quarry (now a parking lot) at Castle Lanes Bowling Alley, locality 17, Figure 27. (Traced from a photograph.)

cut outcrop a path from the end of Russia Street crosses Sunnydale. One can continue down Sunnydale past large blocks of basalt and chert rising out of the hillside above the road cut or follow the path east to the top of the ridge and turn right down the ridge crest to see other blocks, including coarse sandstone with red chert fragments (11; Figure 27), sandy tuff with well-formed feldspar crystals (12), pillow basalt with dark red chert and black coatings of manganese oxide (13), silica-carbonate rock (sc)—magnesium or calcium carbonate seamed with anastomosing veinlets of opal—a product of the hydrothermal alteration of serpentine, and finally a large outcrop of basalt (14).

At the entrance of the golf course is a view eastward of Visitacion Valley, San Francisco Bay, and Candlestick Hill. Cross Sunnydale, and take the path to the rock outcrops on the ridge to the southwest. The northernmost outcrop is sandstone. Circle it, and follow the path to a higher outcrop (15), which is chert. From the top of this outcrop is a fine view northward of the McLaren Park ridge, with the exotic blocks, resistant to erosion, rising as isolated crags from smooth hillsides underlain by soft melange matrix.

Return to the sandstone outcrop and take the path southeast along the ridge crest, past a crag of sandstone on the right and boulders of chert on the left. Beyond these outcrops the trail crosses a saddle and ascends to the north side of a tree-covered ridge. Follow the trail east along the north side of the ridge. John Muir Elementary School, to the north, is built on the floor of an excavation whose banks, not visible from the trail, expose the dark-grey clay of the melange matrix. The large outcrop ahead on the nose of the ridge is a block of green

gabbro with strong gneissic banding (16). It could have been derived from a gabbroic layer of the subducted oceanic crust.

After examining the gabbro, return to the path, and follow it as it bends sharply around the east end of the ridge and climbs west to the hilltop along the south side of the wood. From the hilltop, San Bruno Mountain and the hills in front of it can be seen. Directly below is Geneva Avenue and the quarry of Castle Lanes Bowling Alley. Along and near the path are scattered blocks, mainly of sandstone but also basalt and chert, some rising as low crags out of the grassy hillside, others uncovered by excavation for the trail. The melange matrix is not exposed but will slither into any excavation made by the unwary. Follow the ridge crest west to its termination in a barren field near the southeast corner of the playground fence, go out to Geneva Avenue, and walk east to the pedestrian crossing to the bowling alley (17).

The west wall of the quarry (Figure 28) exposes beds a few centimeters to 5 m thick of massive sandstone that were presumably deposited as submarine sand flows or turbidity currents (See Trip 3, Route A to the beach, for a discussion of turbidity current deposits and the Bouma units and criteria referred to in the next paragraph.)

The sandstone beds (Bouma A units) are massive and without internal laminations, except near their tops, and are coarsest toward their lower (south) sides, where they rest with sharp contact on shale beds. They grade at their tops into thinly laminated, fine sand and silt (Bouma B, C, and D units) and finally into shale (Bouma E unit). On the under surfaces of a few sandstone beds near the south end of the west quarry wall are load casts and flute casts.

The former are described in Trip 3, Route A; the latter are blunt-ended ridge- like projections of the sandstone into the shale, thought to be the casts of hollows scoured into the underlying clay (now shale) by the turbidity current that deposited the sand. In the middle of some sandstone beds are layers of shale pebbles 1–20 cm long, presumably ripped by the turbidity currents from the clay of the seabed over which they flowed. Near the north end of the quarry wall, a 3-m sequence of black shale represents a long period when mud rained onto the sea-floor uninterrupted (locally at least) by massive sand flows.

The coherence and relative lack of deformation (except for some minor faults) of this sequence of sandstone and shale contrasts strongly with the pervasive shearing of sandstone and shale matrix in exposures of the melange north of Geneva Avenue.

After examining the turbidites and, if you need, consuming a cocktail or snack at the bowling alley, walk east on Geneva to a bus stop opposite the end of Brookdale, and board the next eastbound No. 15 (Third Street) bus.

Lines 52 (continued) and 15 (Figure 23)

Those who elected to miss side trip B will travel on Prague, Russia, and Moscow avenues to Geneva Avenue. Ahead on Prague and to the right from Russia, rows of "little boxes" can be seen clinging to the steep slopes of the northern ridges of San Bruno Mountain. At Geneva and Mission, cross to the south side of Geneva west of Mission, where you board the eastbound No. 15 (Third Street) bus.

Geneva is along the contact of the melange belt with the San Bruno Mountain Block. A few blocks east of Mission, Crocker-Amazon Playground is on the left (north) with McLaren Park behind it—fields and groves on the gently rolling hills with scattered craggy outcrops characteristic of melange terrain. The crags are the exotic blocks: schist, basalt, serpentine, chert, sandstone, and gabbro; and the gentle grass-covered slopes are commonly underlain by the dark-grey clay-rich matrix of the melange.

The steep hills on the south are characteristic of topography in the San Bruno Mountain Block. Castle Lanes Bowling Alley, on the south side of Geneva, sits in an abandoned quarry whose walls are superb exposures of the turbidites of the San Bruno Mountain Block. You will have to glance quickly backward to see them. These exposures are the last stop on side trip B, and you can see them by leaving the bus at Brookdale and walking back to the bowling alley. The last three paragraphs of the side trip describe the outcrop.

No. 15 turns left from Geneva onto Santos. The large building to the southeast is the Cow Palace,

home of livestock expositions, rodeos, rock music festivals, circuses, political conventions, and other indoor shows. On the right beyond the turn is an isolated steep hill of sandstone, which may be of the San Bruno Mountain Block.

For a few blocks the bus goes through the pleasant-looking Visitacion Valley public housing project, reminiscent of Monterey Colonial architecture. Note the mural on the left as the bus turns onto Sunnydale. The bus makes several right-angle turns through a working-class residential area built mainly in the 1940's and 1950's. After turning onto Arleta, the bus makes a turn to the left from Arleta onto Bayshore Boulevard, a wide traffic artery. Ask the driver to let you off at the stop one block north of the turn from Arleta (Hester Avenue) to make side trip C.

Side Trip C (Figure 29) Hester Avenue circles a hill located between Bayshore Blvd. and the Freeway. From it is a good view of the south side of Candlestick Hill and the transition between the Twin Peaks Block and the melange belt. At the south junction of Hester and Bayshore, sandstone is exposed in the road cut; east of this, rare exposures of crushed shale mark a broad shear zone. Weathered basalt or diabase is exposed in the road cut at the sharp bend around the hill.

The view from this bend includes San Bruno Mountain, San Francisco Bay, and Candlestick Hill. On the far side of the bay is the Diablo Range, whose core is Franciscan Assemblage (much of it in the blueschist facies, see Trip 6) surrounded with steep fault contacts by coeval rocks of the Great Valley Sequence (see Trip 7) that underlies the foothills of the range. The fault is considered by most to be the deformed Coast Range Thrust, along which the Franciscan was subducted beneath the Great Valley Sequence, but others doubt the existence of the thrust (see *Ernst* [1981] for discussions). The highest point visible is Mt. Hamilton, 1333 m (4372 ft), whose summit is decorated with the white domes of Lick Observatory. The Santa Cruz Mountains are west of the bay. Their near skyline (behind San Bruno Mountain) consists mainly of Tertiary rocks west of the San Andreas Fault, but the more distant (and higher) part of the range, behind the bay, is east of the fault and is Franciscan. The highest point, in the far distance, is Loma Prieta, 1155 m (3791 ft).

Most of the flatland in the foreground was once part of the Bay and was filled with San Francisco's garbage. Part of the flat was once a steep hill of chert, basalt, and sandstone, which was leveled to make the lid on the garbage.

The forested summit of Candlestick Hill is a city park (Bayview Park) accessible from Key Avenue, where this side trip ends. Until the 1950's its round-

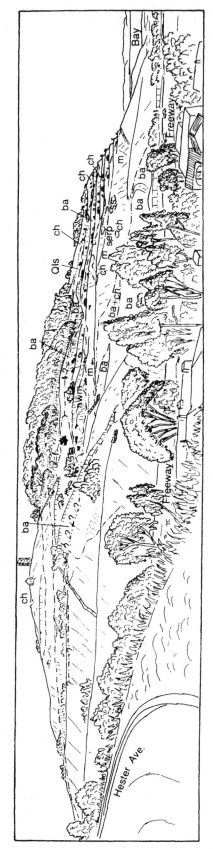

Fig. 29. Panorama of the southwest side of Candlestick Hill from the corner of Hester Avenue: **ba**, basalt; **ch**, chert, **ss**, sandstone (in large block); **m**, melange; **wm**, weathered melange. (Sketched from photographs taken March, 1983.)

ed grassy slides plunged directly into the bay, affording peaceful fishing. The sides were quarried away into the barren terraced slopes we see now (Figure 29) for fill for the Candlestick Park baseball stadium parking area. The loss of the natural landscape has exposed the structure of the hill.

Snetsinger [1979] and *Bonilla* [1971] have mapped this hill in detail, and the interpretation in Figure 29 is based in part on their maps. The crags along the hilltop are chert, as is some of the reddish rock in the upper benches of the cut. Most of the orange-weathering rock is basalt, but some is weathered sandstone of the melange. The grey lower slopes are sheared sandstone and shale, and the light greenish-grey material at the base is a much deformed large block of basalt in the melange. In some seasons, landslides break from the grey slopes. The major break between the Twin Peaks block and the melange may be the base of the basalt near the hilltop, but basalt near the bottom of the cut and on Hester Street, and the former basalt and chert hill a few blocks to the south, show that the contact is gradational, being a southward increase in the proportion of sheared rock and a decrease in the size of the coherent blocks.

Walk north on Hester to Third, and cross the bridge over the freeway on the east side of Third. You can observe through the cyclone fence the sheared rock of the melange and, in it, lenticular blocks of sandstone and basalt. Some blocks contain thin beds of shale. In general, both blocks and shearing dip northeast. Red tones along the hilltop mark chert and weathered basalt. The crumbly orange rock in the first accessible outcrop north of the fence is weathered basalt. A few feet farther are obscure exposures of bedded chert. The view up Le Conte is to crags of chert near the hilltop. Reboard bus No. 15 at the south side of Key Avenue (across from Saint Paul of the Shipwreck School).

The 15 Line (continued) and the 42 Line (Figure 23)

Hunters Point, the hill to the north on the right (east of Third Street), is serpentine. Its grey-green road cut exposures can be seen from many points along Third. The hill to the west is sandstone but lacks visible exposures. Near Palau, watch for explosively colorful murals on the walls of Hunters Point Community Center on the right, and just beyond, also on the right, facing north on Newcomb, is the South San Francisco Opera House, built in 1888 in Italianate style.

Beyond, Third Street enters the major industrial area of San Francisco. India Basin Industrial Park, site of former abattoirs, is on the right, and beyond, the drawbridge over Islais Creek Channel. To the west (left) beyond the elevated freeway is Bernal Heights, an erosional remnant of a coherent slab consisting of chert depositionally on basalt with a nearly flat contact. The basalt in turn rests on a melange matrix of sheared sandstone and shale which once probably encased the entire slab. North of Bernal Heights and Islais Creek, the high grey hill behind the huge gas tank is Potrero Hill, about 100 m high, the largest exposure of serpentine in the city.

Beyond Army Street, Third Street begins to cross numbered streets, from 26th down to 16th. Between 23rd and 20th, it crosses the west edge of the site of a steep serpentine hill that was leveled for a factory district, and at 20th, on the left, behind a bank building, outcrops and boulders of serpentine landscape a parking lot. The red factory buildings on the right were formerly the American Can Company but are now rented to small manufacturing firms and as studios. Behind them is the Potrero steam power station.

Beyond 16th, Third Street crosses filled land that was once Mission Bay. The bus veers left onto Fourth and crosses China Basin (all that remains of the bay). Two blocks beyond, at the corner of Fourth and Townsend, is the transfer point to the No. 42 bus (Downtown Loop).

Take the Red Arrow No. 42 bus on the southwest corner of Townsend and Third (beside the railroad station) east on Townsend. The bus goes through an industrial district with a maritime flavor, then turns left onto Second Street and over the west flank of Rincon Hill, a small hill of sandstone of the Nob Hill Block (V, Figure 5). In the 1850's the wealthy lived here, away from the fog, mud, and sand. It is now overwhelmed by the Bay Bridge approach. The 42 bus turns right from Second onto Folsom, left onto Fremont, and passes the East Bay Terminal (EBT, Figure 23) between Howard and Mission (AC Transit buses to Berkeley and the East Bay have splendid views from the bridge). No. 42 crosses Market onto Front, turns left onto Pine, and right onto Sansome. The Ferry Building can be seen at the foot of Market on the right. The California Division of Mines and Geology at present has a museum, library, and publication sales office on the second floor, open 8–5, Monday and Friday. *May be moved!*

At the corner of Pine and Sansome you are in the heart of the financial district. Banks abound. Between Clay and Washington, on the left the Transamerica Pyramid (T) juts into the sky. Between Washington and Jackson, the U.S. Appraiser's building on the right houses the U.S. Forest Service and other federal agencies. East of it on the same block, on the fifth floor of the U.S. Customs House (C), is the Public Inquiries Office of the U.S. Geological Survey (USGS), where USGS map and book publications and U.S. Coast and Geodetic Survey nautical charts may be purchased.

On the left, beyond Broadway, are the steep cliffs of Telegraph Hill (TH), the outcropping edge of a massive sandstone bed. The cliff is the back wall of a quarry that provided rip-rap for the Embarcadero. Side trip D is a climb up the steps of this cliff to see the sandstone close at hand and visit Coit Tower at the top.

Bus No. 42 continues northwest on the Embarcadero and left onto Bay. From Bay it turns right onto Columbus and left on North Point to Van Ness. Fishermen's Wharf is three blocks north of Bay and Columbus, and on North Point, No. 42 passes The Cannery and Ghirardelli Square. At Van Ness you are beside Fort Mason (FM) (Trip 2). No. 42 continues south on Van Ness to your point of origin.

Side Trip D: The Sandstone of Telegraph Hill
Alight from the No. 42 bus as near as possible to Filbert Street, and walk west on Filbert. The Filbert steps here climb the east face of Telegraph Hill. The cliff appears to be a single massive turbidite bed, possibly more than 100 m thick. However, wisps of fine-grained material in the cliff alongside the steps suggest that the apparently single deposit may be the product of several turbidity currents, each of which eroded the underlying Bouma B through E units before depositing its burden of sand. These wisps of fine-grained material also indicate that the sandstone dips west. It is interpreted by *Schlocker* [1974] to be the east limb of a syncline whose hinge line runs down Columbus Avenue.

At the top of the first flight of steps is a delightful garden walkway with two boardwalk lanes branching to the north (Napier Lane and Darrell Place). The houses here predate the fire and earthquake, and some date from the late 1850's. Early in this century this area was favored by struggling artists and writers because it has a fine climate and view and was cheap. Once the affluent realized what the artists had found, rents rose, and struggling artists no longer live here. These houses were saved from the 1906 fire by the quick thinking of their inhabitants, who soaked cloth in wine and spread it over the roofs to put out the sparks that were raining from the conflagration to the southwest (the water supply had been knocked out by the earthquake).

Cross Montgomery Street (here split into two levels) and climb the steps to Telegraph Hill Way. Turn right on Telegraph Hill Way to the circular plaza in front of Coit Tower. The jasper stones along the walk are decorative imports from jasperized chert or basalt elsewhere in the Franciscan.

Coit Tower is a gift of Lillie Hitchcock Coit, a col-

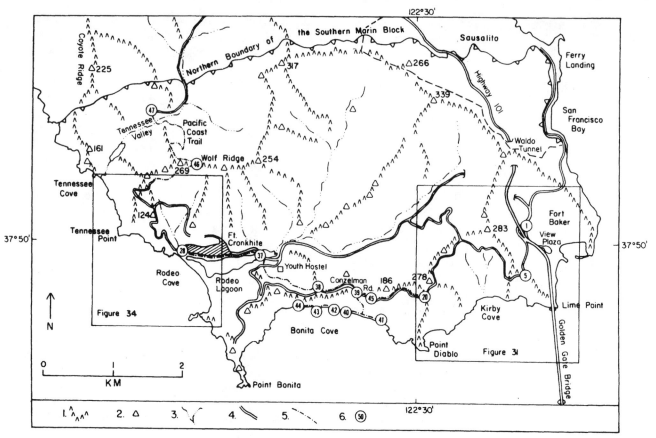

Fig. 30. Outline map of the Southern Marin Block (see Figure 6 for location), showing areas visited in Trips 5a and 5b: **1,** ridge crest; **2,** summit, **3,** stream or valley bottom; **4,** road; **5,** route of walking trip between Fort Cronkhite and Kirby Cove and Tennessee Valley Road; **6,** locality mentioned in text.

orful citizen of 19th-century San Francisco, who in her childhood was the mascot of a volunteer fire-fighting company and all her life had a deep affection for volunteer fire fighters. She left money for both a statue at the northwest corner of Washington Square and this tower as their memorial. Its shape is supposed to be vaguely reminiscent of the nozzle of a fire hose.

The view from the top of the tower is spectacular (see Figure 16 for part of it), and the murals on the ground floor, painted during the late 1930's by artists of the WPA Arts Project, depict social and economic life during the Great Depression.

You can return to Sansome Street and continue the trip as described, or take the No. 39 (Coit) bus from Coit Tower to Washington Square and Columbus Avenue, walk northwest on Columbus and north on Taylor to Bay to rejoin the No. 42 bus there.

Trip 5. Marin Headlands: Pillow Basalt and Chert

This trip is in two parts: 5a and 5b. The sea cliffs and road cuts of the Southern Marin Block (Figure 30) have some of the finest exposures of pillows and

minor folds in chert to be found anywhere. The pillow exposures rival those along the midoceanic ridges, so if you were unable to book passage on the Alvin, take Trip 5b. Trip 5b can be made easily only during Sundays and holidays in winter and only on weekends and holidays in summer. The paper by *Bedrossian* [1974] has a colored geologic map and is a useful supplement. As with Trip 3, wear scuffable clothes, hiking or tennis shoes, and carry a sweater, camera, and lunch, but no hammer. This is a national park. *Do not* wear shoes with smooth or slippery soles.

Two bus lines serve these trips: (1) Golden Gate Transit bus 10 starts at the East Bay Terminal and makes stops at 7th just north of Market and on Van Ness at Geary, Sutter, Clay, and Union streets. (2) Muni Bus Number 76 runs on weekends and holidays June 1 to November 1 and on Sundays only November 1 to June 1 from Caltrans depot at 4th and Townsend via the East Bay Terminal and Sutter and Van Ness to (28). Call 673-MUNI for schedules, and ask driver for return stops.

Figures 30, 31, and 34 are maps of these trips. Localities (1) through (27) are on Figure 31, (28) through (36) on Figure 34, and (37) through (48) on Figure 30.

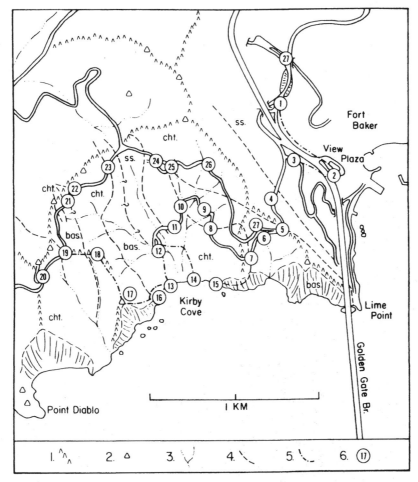

Fig. 31. Outline map of Kirby Cove, showing localities visited on Trip 5a: **1,** ridge crest; **2,** summit; **3,** valley or ravine bottom; **4,** contact between bedrock types (generalized): **bas,** basalt; **cht,** chert; **ss,** sandstone; **5,** route of walking Trip 5a (where not on roads); **6,** locality mentioned in text. Number 27 in northeast corner of map is the southbound Golden Gate Transit stop.

Trip 5a

Trip 5a is to Kirby Cove (Figure 31) at the forested valley just west of the Golden Gate Bridge. The mountains on either side of the cove are carved in southwest-dipping slabs of chert and basalt; the cove, valley, and pass to the northwest are apparently eroded along sandstone and shale, probably in fault contact with the rocks on either side. Most ridge crests here (and elsewhere in Southern Marin Block) are the outcropping edges of chert units that are resistant to erosion because chert is not chemically weathered. Basalt underlies side slopes and saddles, sandstone and shale underlie the valleys and low passes. Unweathered sandstone and basalt form sea cliffs as prominent as those in chert.

The ridge northeast of Kirby Cove is held up mainly by a slab of chert about 230 m thick near the head of the valley but thinning to about 30 m at the north tower of the bridge. It is in fault contact with sandstone on the northeast. A wedge of basalt about 230 m thick makes up the dark-green cliffs between the bridge and Kirby Cove and thins northwestward

to about 60 m where it crosses the ridge east of the cove and to about 10 m at the head of Kirby Cove valley. Above this basalt, remnants of a slab of chert cap the nose of the spur on the east side of the cove and the ridge that splits the head of the valley. The road to Kirby Cove is alternately on the basalt and the overlying chert.

If you take Golden Gate Transit buses 10 or 20, get off at the first stop north of Golden Gate Bridge (1), on the Alexander Avenue off ramp. Muni bus Number 76 may go via the headland road and you can get off at (5); otherwise, leave it also at (1). Walk back along the east side of the freeway to the View Plaza (toilets and drinking water). On the west side of the plaza parking lot (2), take steps down to a pedestrian underpass beneath the bridge and the road uphill from the west side of the parking area west of the bridge. Road cut (3) exposes sandstone.

Turn left onto the road leading uphill into the headlands. First outcrops are sandstone. Road cuts beyond expose only landslide debris for 300 m. At (4) is chert west of the fault. The thin beds of chert and shale are in tight chevron folds with wavelengths and amplitudes of 2–4 m. Chert extends to (5) but is not tightly folded.

The eastern contact of the wedge-shaped body of basalt is just beyond the sharp curve at (5), complicated by a repetition of chert and basalt. The easternmost basalt exposures are thoroughly weathered.

Just beyond (5), take the dirt road downhill to Kirby Cove. Basalt in the first road cut is brown and punky, but 30 m west it is green and hard, and pillows are evident. At (6) a rock outcrop about 9 m high and 6 m below the road has pillows exposed in three dimensions on its east face. Pillows can be recognized in cross section along the road beyond. At (7) the road encounters the chert slab that caps the spur facing Kirby Cove and that dips about 30° southwest.

Beyond (8) the road crosses back onto basalt, in which it remains to the head of the valley. Rounded upper surfaces of pillows are exposed at (9), and around the next curve, pillows are exposed in cross section. At the head of the valley the road turns abruptly south and reenters the chert (10). At the next curve to the right (11) the road cut exposes a single, bedding surface folded into a sharp south-plunging syncline and anticline. West, near the next bend to the left, are obscure exposures of weathered sandstone.

The forests of eucalyptus, Monterey (*radiata*) pine, and Monterey (*Macrocarpa*) cypress were planted here by the Army for camouflage when this was a fort. Now the GGNRA has several group camps in the forest (toilets and water; telephone for reservations up to 2 months in advance). The road ends in a loop with parking area (12). South from the loop a fire trail follows the west side of the valley to the beach, and a path through the forest from the parking area goes past toilets and faucets to the east side of the valley and the beach.

Across the mouth of the valley is an abandoned shore battery disguised to appear, from the sea, as a line of dunes. At (13) and (14) the paths to the beach are down steps in this artificial landscape. The shore battery rests on deeply weathered colluvium that rests, in turn, on deep soil and weathered bedrock, which descends beneath sea level, indicating a former protracted low stand of the sea (see trip 5b for further conclusions).

At the east end of the beach, tightly folded chert has two sets of minor folds: an early set, whose hinges trend S 75°W, is bent around a later set of more open folds whose hinges trend almost due north. These are best seen on the wave-cut platform exposed at low tide in late winter but are evident in the cliff at any time. The chert-basalt contact of (7), (8), and (10) reaches sea level at (15), just beyond a rocky point. At low tide, one can wade around the point to a deep sea cave eroded along the contact.

Dark-green cliffs of basalt are visible beyond, and the rock offshore consists of basalt with a capping of chert.

The cliff at the west end of the beach is pillow basalt. Near exposures are deeply weathered and bright orange, but farther west, cross sections of pillows are evident in less-weathered basalt. The best pillow exposures are at (16), at the west end of the cove; pillow tops are clearly to the southwest. About halfway along the cliff, a belt of sheared volcanic breccia with thin lenses of chert probably marks an interval of long duration between two eruptions.

Those with pressing business in the city can return via the Kirby Cove road to the View Plaza. From here there are two ways home, both having psychological hazards: (a) return north along the east side of the freeway and Alexander Avenue past the northbound bus stop to a southbound Golden Gate Transit or Muni Number 76 bus stop at the road junction just north of the large road cut in chert and basalt ((27) in northeast corner of Figure 18), and take any southbound Golden Gate Transit bus. (Alternatively, where you first got off, you can take the next northbound bus to Sausalito, and the Sausalito Ferry (bar on board) to the foot of Market Street.) (b) Walk south across the Golden Gate Bridge (sidewalk on east side; 3 km), down steps at the toll plaza, and take Muni bus 29 (see Trip 3 for details).

The energetic and adventurous with more time to spare may wish to climb the steep trail through the forest along the cliff top west of the cove to the Headlands road 245 m (vertically) above to see fine road cut exposures of folded chert. This is a hazardous route and should only be taken uphill. Street shoes or other slippery footgear should not be worn.

The well-beaten path goes straight up the slope for 20–30 m, diagonals to the left, and crosses the head of a landslide scar at the top of the cliff above the west end of the cove, where it is nearly level across the east-facing forested slope, to reach the cliff top. Here it crosses the southwest-dipping contact between basalt and the overlying chert (Figure 32). The path continues straight up the slope with a sheltering screen of trees between it and the cliff top. The safest route at the top of the forest is to the right, over the branches of a toyon tree, where you come out onto a cliff top. Here you look straight down about 100 m, past cliffs of ribbon chert, to the sea below. The path continues upward along a sharp ridge crest; the cliff is on the left. Avoid touching poison oak.

At the highest point of the cliff, the path turns inland along the ridge crest, and about 30 m from the cliff top, in the first chert outcrops, it passes the brass marker of Triangulation Station Gate (17).

Here, bedding locally strikes east and is nearly vertical, but the chert is actually isoclinally folded about west-plunging fold hinges, one of which is exposed on the east face of the outcrop, about 2 m northeast of the brass marker, and another in the outcrop 6 m north. Farther north the chert lacks the tight minor folds and strikes N 25°–40°W, parallel to the ridge and path, and dips 30°–35° southwest.

At (18) the path turns abruptly west straight up the slope. The chert bed is marked to the north by a line of prominent crags. Poorly installed culverts caused deep gullies below the road to the north. The path up the steep slope is partly over poorly exposed chert but mainly over deeply weathered basalt, whose contact with the next chert above is exposed at the sharp bend (19) where the path joins the road. Minor chevron folds whose hinge lines plunge 20°–30°SW are exposed in the road cut to the west.

This point is a favorite for sightseers and photographers. Along the bluff south of the bridge, across the Golden Gate, are serpentine and melange. The high, forested plateau due south across the Gate is underlain by chert, basalt, and sandstone like that here, but it is more chaotically deformed and dips generally northeast. From hilltop view sites 30 m higher is a 360° panorama. Toilets are located at (20).

To return to San Francisco, walk north from (19) on the road past road cuts in pillow basalt (see end of Trip 5b for a day-long alternative). Where the road curves east is another body of chert that overlies basalt to the east, which in turn overlies a 2–3 m bed of sandstone resting on chert and folded into a shallow anticline (21). East of (22), chert in road cuts is folded into tight chevron folds plunging

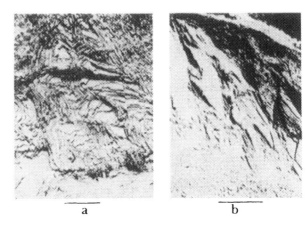

Fig. 33. Minor folds in chert at locality 26, Figure 31, looking northeast: (*a*) folds with two sets of hinge lines exposed in cross section at locality 26; (*b*) folded bedding surface 50 m south of locality 26.

20° S70°W. Chert continues to (23), beyond which is basalt almost to the intersection at the pass.

Keep right at intersection. For 100 m the road is in sandstone. Beyond is chert, with a 10–20 m layer of basalt exposed on both sides of the next spur (24 and 25). The basalt must strike into the sandstone, proving the sandstone-chert contact a fault. This thin layer of basalt is the northwest extension of the body between the bridge and Kirby Cove.

At (26) is a photogenic outcrop of multiply folded chert (Figure 33a). The next road cut south has cleaned off one tightly folded bedding surface, showing the folding, which plunges 40°–45° S30°W, to be anything but cylindroidal.

South of (26), colluvium-filled gullies, exposed in cross section, testify to a late Pleistocene period of deep erosion followed by a long period of erosional stability while colluvium accumulated and was weathered. About 100 m farther south the road enters the wedge of basalt, which is exposed for 150 m, almost to the curve, where the road crosses onto the overlying chert. Basalt is encountered just past the next sharp curve, and road cuts across the head of the swale (27) expose dark-green unweathered basalt where well-preserved pillows show that the basalt is right side up and the present direction of the slope of the seafloor on which the pillows accumulated is to the southwest. Just beyond is the junction with the Kirby Cove road. Instructions are given above for the trip home.

Trip 5b

Trip 5b can be made easily only by Muni Bus Number 76 (provided the National Park Service continues to subsidize this service). It is one of the great geological travel bargains around (60 cents each way, as of this writing).

The bus terminates at (28), (Figure 34), where

Fig. 32. Cliffs and mountain west of Kirby Cove from near locality 7, Figure 31, looking west. The trail from locality 13 to locality 19 goes along the south edge of the forest near the cliff top in the left corner of the picture and thence along the nearer of the two ridges. Locality 17 is at the first chert outcrop north of the lone tree at the highest point of the near sea cliff.

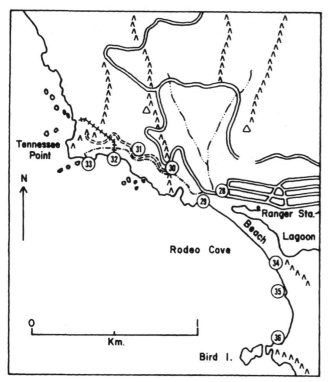

Fig. 34. Outline map of Rodeo Cove, Tennessee Point, and vicinity, showing localities visited in Trip 5b. (See Figure 30 for location; see Figure 31 for explanation of symbols.)

there are toilets and drinking fountains; the ranger station, one block east, sells popular geologic and other reports and gives trail information.

The Franciscan here (Figure 34) strikes east and dips south. A thick chert unit makes Wolf Ridge (Figure 30) the skyline to the north, and a second unit is responsible for the chert quarries in the hill about 1 km east as well as the two high hills on either side of the valley north of (28). Between these two chert units is a complex belt of basalt, sandstone, and minor chert beds plus melange, which forms a line of saddles. The basalt of Tennessee Point, a goal of this trip, is a wedge-shaped body of this belt pinching out to the east. The valley of Rodeo Lagoon is probably underlain by sandstone, at least at its lower end; this sandstone is exposed in the first outcrops on the walk. The low plateau south of the lagoon has sandstone and melange at its north end and farther south is underlain by basalt. The erosional resistance of a north-trending diabase dike intruded into the basalt may be responsible for the narrow promontory of Point Bonita (Figure 30).

At the north end of the beach (29) is coarse, dark sandstone, graded and cross bedded, with clay chips and volcanic rock fragments. A thick reddish Pleistocene soil, on which colluvium (now deeply weathered) was deposited, descends below sea level here.

A younger, clay-rich soil, with dark grey humus at top, caps the colluvium. These soils indicate that, during the later Pleistocene, sea level here has not been higher with respect to the present land surface, than now.

The route to Tennessee Point goes up the steps between the road and the sea cliff, thence north along the cliff, past the head of a narrow cove eroded along a N30°E-trending fault zone visible from cliff tops west of the cove. At (30) the route joins a narrow dirt road branching from the paved road to the ridge crest. An enigmatic body of pillow basalt is exposed at this junction, but road cuts beyond expose 1–2 m beds graded from granules at base (north) to coarse sandstone at top (south). Around the next spur the road is at the cliff top, and Tennessee Point is ahead, with dark-green to black basalt along the shore. Nearby below, the cliff is a bedding surface in red and green chert.

At (31) a path worn in the grass branches to the left and leads to a hole in the fence at the cliff edge. The area beyond this fence is still used occasionally as an ammunition detonation area and is posted "Keep Out." I have to advise you to go no further. However, if you choose to ignore my advice, do as I do, and go through the hole in the fence; once inside the detonation area, stay on the beaten path.

If you look back toward (32) from the path, you will see chevron folds with amplitudes of 2–6 m in the chert sea cliff.

Where the path reaches the floor of a south-trending valley across Tennessee Point, a narrow slot leads down to the shore platform (33). This platform is carved on well-exposed pillows whose plane of flattening strikes about east and dips about 40°S. Molding of pillows around the backs of older pillows, and the direction of attachment to their feeder tubes, indicate that younger rocks and the paleoslope are both to the south. Well-developed radial shrinkage cracks extend 10–15 cm into the pillows, forming polygonal columns 5 cm on a side.

On a field trip in 1980, Tanya Atwater pointed out some rare features here that show precisely the present inclination of a surface that was perfectly level when the pillows were formed. These are flat-based cavities—now filled with quartz and calcite—in some pillows at the west side of the bench. Thin plates of basalt parallel to the flat floors extend into these cavities from their sides. She had seen these features in newly erupted pillows during submersible dives to the midoceanic ridges, and she recognized that they result from lava draining from a newly formed pillow, creating a void that is instantly filled with seawater breaking at high pressure through the chilled pillow rim. The cold seawater chills the level top of the liquid lava still in the pillow, freezing in the paleolevel. Repeated withdrawal of lava leaves behind chilled paleolevels as the thin plates. Because pillows can form like candle drippings on slopes of almost any inclination, even up to vertical, as *Fornari et al.* [1978] discovered off the southest coast of Hawaii, these rare features are important for many studies of this basalt, including paleomagnetic studies to determine the latitude at which the pillows were formed.

The thick mass of pillow basalt that makes up Tennessee Point is in fault contact with sandstone and shale to the north and is probably overlain conformably by the chert to the south. It wedges out to the east near the top of the coastal ridge, suggesting the flank of a submarine volcano. The flat cavity bases and flattening of the pillows suggest that this seamount had gentle slopes.

As you return up the slot to the path, notice the change in color and consistency of the basalt through the weathering profile. Return exactly the way you came, at least as far as the fence, and thence back to the bus stop (28).

Rodeo Beach is one of the few deposits in the world dominated by material in the 1–4 mm size range. This is because of the predominance of chert in its source area [*Wakeley*, 1970]. Carnelians (semiprecious translucent red-to-orange chalcedony) eroded from vesicle fillings in basalt can be found among the beach granules.

At the south end of Rodeo Beach (34–36, Figure 21) are exposures of pillow basalt and diabase, and also subaerial Pleistocene deposits that indicate that this is a currently subsiding stretch of coast. The northernmost 200 m of outcrop (34) consists of four south-dipping sequences in fault contact with each other and which are locally complexly deformed. The northernmost is chert. South of this for 30 m is a many-times-repeated sequence of medium-to-coarse massive sandstone grading upward to thinly laminated fine sand and silt with abundant coalified plant remains on partings. South of this is mixed pillow basalt and chert, apparently near the margin of a submarine flow. Still farther south, across a landslide that brings large sandstone blocks to the beach and apparently marks a narrow melange zone, is pillow basalt. This last unit forms offshore pinnacles (35) and a promontory that blocks access to the cove to the south, except at low tide. However, one can climb the hill at the south end of the lagoon and reach this cove via trails through the sand dunes.

South of this point, and past a N20°W-trending vertical fault zone marked by blocks of sandstone, the sea cliff is medium to coarsely crystalline diabase intrusive into the basalt. The north-trending contact is at the base of the cliff, and pillows are well exposed in cross section on the wave-cut bench below the cliff. Their tops are to the east. A branch of the

diabase dike is on one of the offshore rocks accessible at low tide.

South, at the back of the cove, is a deposit of nearly flat-lying Pleistocene alluvium—alternating beds of silt and fine sand with scattered angular chert fragments. The silt and sand were windblown into this cove from the glacial-age floodplain of the Sacramento River during the minus 100 m sea levels at the maxima of glaciation. The alluvium rests on a thick soil developed on deeply weathered diabase and basalt. Elsewhere along the California shoreline, as at Santa Cruz, Bolinas, and Sonoma County, uplifted wave-cut benches backed by fossil sea cliffs are evidence that during the Quaternary the land was uplifted as much as 250 m. These easily eroded deposits, in a sea cliff facing the open ocean and subject to wave attack, are evidence that San Francisco and southern Marin County are subsiding in relation to other parts of California. Similar deposits are at (13), (14), (29), (1, Trip 3) and (6, Trip 6). The sagging of the crust here may be due to a slight pulling apart of the two sides of the San Andreas Fault, which here makes a slight bend to the right.

The north-facing cliff at the south end of the cove has intrusive diabase at its east end. Westward are submarine volcanic breccias and pillow basalt. A thin, purple bed capping a mound of pillow basalt at (36) is limestone and chert, which were deposited during an interval between eruptions, and probably much closer to the equator than here.

From (28) one can return to San Francisco via Muni Bus 76 (if it is running) or a hike back along the ridge just north of the Golden Gate to bus stops near the Golden Gate Bridge. On the way, visit Bonita Cove. To do this, hike east along the road from (28) and across a trail bridge at the head of the lagoon (37, Figure 30). Follow the Pacific Coast Trail (marked by signs), up to the hill and along the south side of the wooded valley housing the Headlands Hostel, to the saddle (38); go a few meters south along the ridge to the ridge top road, and follow the road east onto pavement. The road cuts have obscure exposures of sandstone, chert, and basalt. Where the road crosses to the south side of the ridge is a parking lot (39) and a trail to Bonita Cove beach. This beach is replete with interesting exposures, especially on a warm sunny day.

The rocks along Bonita Cove strike east and dip south; pillows and all sedimentary structures indicate that younger rocks are to the south (unless the rock sequence is repeated by faults, which is probably the case). Midway along the shore, a sharp point with a spirelike crest (40) that blocks passage between the two halves of the beach has spectacular pillows (Figure 35). Much of the bluff behind the beach consists of sandstone resting on chert. A ver-

44

Fig. 35. Pillow on shore platform on beach at Bonita Cove (locality 40, Figure 30). (Photograph by David Bice.)

tical to south-dipping fault lies between the sandstone and the pillow-basalt and chert that make the headlands to the south such as at (40) and Diablo Point.

The trail to the beach branches 100 m south of (39). The east branch goes to the pillow-basalt locality (40) and to exposures of the depositional contact of sandstone on chert (41), where angular fragments of chert embedded in the basal sandstone indicate that the chert was already segregated into layers and was considerably consolidated when the overlying sandstone was deposited. At (40) erosion of the matrix has exposed pillow tops in great detail (Figure 35). The pillows appear to have flowed south, and their feeder tubes appear to have collapsed along tiny faults as the lava drained into their lobes. The platform on which the pillows are exposed is accessible only when the tide is lower than +0.5 m (at the Fort Point Gage; check tide tables and calculate the proper time for a visit). About 50 m east of (40), blocks of pillow basalt fallen from the cliff onto the beach have abundant interstitial, coarsely crystalline pink limestone. Although access via the east branch trail is easy, dense stands of poison oak border the trail, and you should avoid touching any vegetation.

The west branch of the trail is to well-cemented, well-graded grey chert-pebble conglomerate (43), whose pebbles might be eroded from earlier subduction zone complexes, and to sandstone that is both thinly laminated, cross-bedded, and graded, all at the same outcrop (44). At a sea cave on the east side of the cliff that blocks passage west (at 44), irregular masses of conglomerate within the sandstone have pebbles of metamorphosed basalt, red and green chert, and sandstone, implying reworking of a Franciscanlike complex. Halfway between (43) and (44), cross-bedded sandstone is exposed resting on basalt with about 0.5–1 m of chert at the contact.

To reach the beach via the west branch trail, keep to the uppermost of the many branches of this fork, staying along the cliff tops for about 200 m west of (40); at this point the trail drops steeply and somewhat hazardously to the beach, over a promontory of chert. If the tide is low, return from (43) and (44) along the beach to the west side of (40), and climb the cliff along a notch eroded along a fault zone between basalt and chert. This fault marks the position of the large bodies of sandstone to east and west along the beach that are faulted out here.

From the trail head, follow the ridge crest road east to the hilltop at (20), where you can follow the return route of Trip 5a. The first road cuts encountered east of the beach access are intensely folded and locally faulted chert, probably the same chert unit as that beneath the sandstone on the beach. Next are exposures of pillow basalt, which probably underlies the chert. At (45), deep gullies, cut into the basalt and filled with colluvium to form a smooth hillside, are exposed in cross section in the road cut. A moderately well-developed, clay-rich soil has developed in the colluvium and stands out as resistant ledges in the road cut. It indicates that this steep hillside has undergone neither erosion nor deposition for at least a few thousand years.

To make Trip 5b, or any part of it, on foot (when Muni Bus 76 is not running), follow this ridge crest road west from (19) and (20) and reverse Trip 5b to (28). You can either return by the route you took, or follow the Pacific Coast Trail over Wolf Ridge (46)—altitude 255 m—to Tennessee Valley (47) and northeast along the Tennessee Valley Road to a bus stop at Tamalpais Valley Junction, about 200 m north of where the Tennessee Valley Road joins Highway 1. There you can catch Golden Gate Transit Bus 10 back to San Francisco (or any other Golden Gate Transit bus that passes by headed for San Francisco).

The total length of the hike from the north end of Golden Gate Bridge is approximately 12 miles (19.3 km). The youth hostel accepts guests of any age and makes this possible as a two-day trip. Call 415-728-7177 for reservations.

Trip 6. A Boat Trip to the Blueschist Facies: Angel Island and the Metamorphosed Franciscan

Metamorphic rocks of the so-called blueschist facies are especially important in plate tectonics because they are thought to have been produced only in subduction zones and, therefore, to mark the location of ancient subduction zones that are no longer active. The blueschist facies is named for the characteristic blue color of its amphibole-bearing schists. It forms under conditions of high pressure and relatively low temperature. The minerals of the blueschist facies are thought to have formed at pressures of 5 to more than 10 kbar, corresponding to the weight of 18 to 50 km of rock, but at temperatures of only 100°–300°C [Bailey et al., 1964, p. 110]. At the normal rate of increase of temperature with depth in the earth, the temperature should be in excess of 300°C at these depths. Therefore, it is thought that cold, ocean-floor material was carried to great depths on a subducting slab, the blueschist facies minerals being crystallized at high pressure and low temperature. The theory requires that the rocks be returned to the surface before they were heated to temperatures above 300°C, at which temperatures these high-pressure minerals would break down. The chief minerals characterizing the blueschist facies are glaucophane, lawsonite, jadeite, and aragonite.

A most convenient place to see the blueschists and their relations to other rocks is Angel Island in San Francisco Bay (Figure 6, see also Figures 36 and 37)—an erosional remnant of a slab of Franciscan rocks metamorphosed to the blueschist facies. This slab is more than 245 m thick and rests on unmeta-

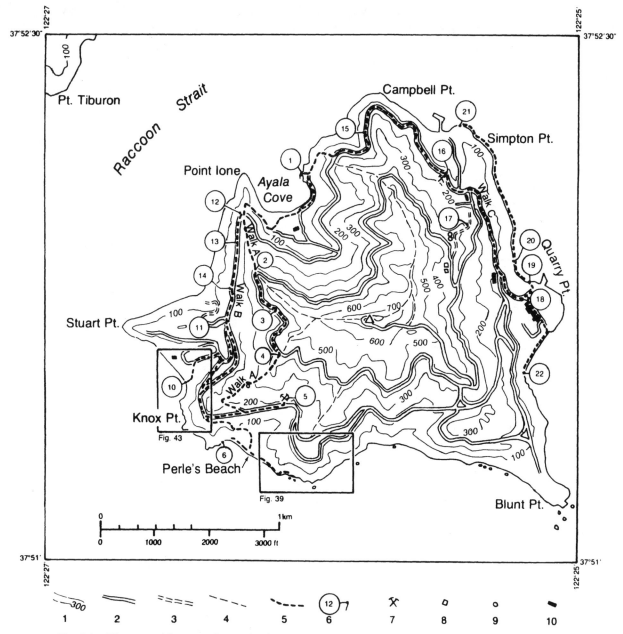

Fig. 36. Topographic map of Angel Island, showing routes of walks, numbered localities mentioned in text, and locations of Figures 39 and 43; contour interval, 100 feet (30.5 m); redrawn from U. S. Geological Survey topographic map of San Francisco North (7.5′) Quadrangle: **1,** contour; **2,** road; **3,** abandoned road; **4,** trail; **5,** route of walk; **6,** locality mentioned in text; **7,** quarry; **8,** reservoir; **9,** water tank; **10,** landmark building. Triangle marks summit of Mt. Carolyn Livermore.

morphosed Franciscan rocks on a flat thrust fault exposed on the east shore of the island. The internal structure of the slab on Angel Island is complicated (see Figure 37). A northwest-trending dike of serpentine separates a northwest-dipping succession of pillow-basalt flows on the southwest corner of the island from an area predominantly of sandstone folded roughly into a broad northwest-plunging syncline and making up the rest of the island. An enigmatic arcuate belt of basalt lies near the middle of the sandstone. All the rocks above the thrust fault have evidence of blueschist-facies metamor-

phism. However, the schists, i.e., the wholly recrystallized rock made up almost entirely of new minerals, occur only as small bodies, generally along contacts of serpentine and basalt, or within the basalt. Although most of the sandstone exhibits a strong schistosity (that is, a tendency to break into thin sheets), this schistosity is due to a flattening of sand grains and to the development of fine parallel fractures. In most of the sandstone and basalt the blueschist-facies minerals are present only as microscopic crystals, partly or completely replacing the original minerals of the basalt and sandstone; they can be

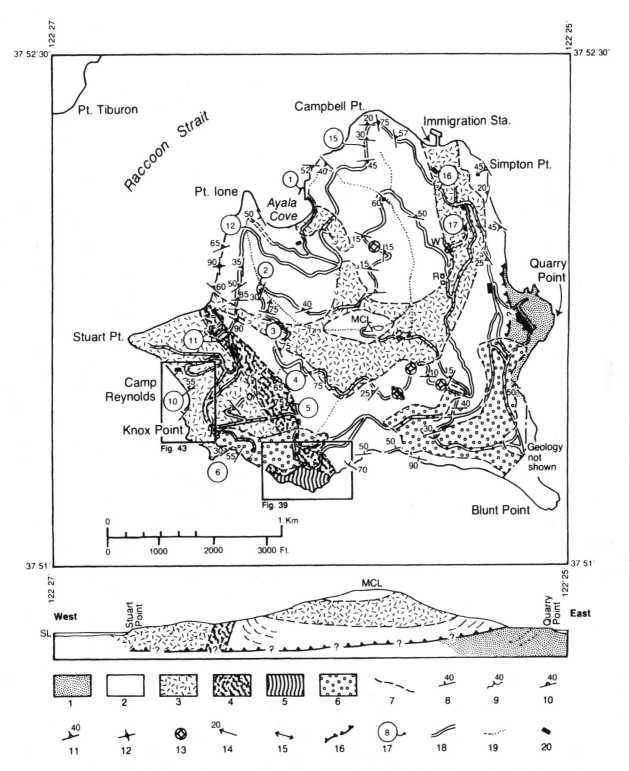

Fig. 37. Geologic map and cross section of Angel Island, showing localities mentioned in text and locations of Figures 39 and 43 (based in part on original surveys and in part on *Schlocker* [1974, plate I]): **1**, unmetamorphosed sandstone of Quarry Point; **2**, schistose sandstone and shale (textural zone II); **3**, basalt; **4**, serpentine; **5**, schist (textural zone III); **6**, Colma Formation; **7**, contact (queried where uncertain); **8**, strike and dip of bedding; **9**, average strike and dip of irregular bedding; **10**, strike and dip of schistosity; **11**, strike and dip of parallel bedding and schistosity; **12**, strike of parallel vertical bedding and schistosity; **13**, horizontal schistosity; **14**, bearing and plunge of linear elongation in plane of schistosity; **15**, horizontal linear elongation; **16**, outcrop of thrust fault and associated melange (also inferred position in cross section), teeth on side of upper plate; **17**, locality mentioned in text; **18**, road; **19**, trail; **20**, prominent building; **MCL**, Mt. Carolyn Livermore; **R**, reservoir; **WT**, water tank.

recognized only through the study of transparently thin slices of rock with a polarizing microscope. Because the plate-tectonic importance of the rocks depends on these minerals, the discussion to follow necessarily refers to microscopic features that cannot be seen on the trip.

Angel Island was first mapped geologically by *Ransome* [1894], who thought that the presence of the schist in small patches along the contacts of basalt and serpentine with sandstone proved that it was the product of contact metamorphism of the sandstone by the hot magmas that later crystallized (he believed) to form the basalt and serpentine. *Bloxam* [1960] restudied the island and showed that jadeitic pyroxene and other metamorphic minerals are widespread throughout both the sandstone and the basalt and that therefore the metamorphism is regional in nature, not contact metamorphism. The island was remapped by Schlocker and Bonilla [*Schlocker et al.*, 1958], and the mineralogy and petrology of the schists was described in great detail by *Schlocker* [1974]. *Coleman* [1965] studied the jadeitic pyroxene and confirmed the presence of aragonite (see also *Coleman and Lee* [1962] for a discussion of the significance of the aragonite).

Three walks are given below to various parts of the island: walk A is to complexly deformed schist adjacent to serpentine along the south shore; walk B is to schist and metamorphosed chert within the basalt sequence of the west shore; and walk C is to the thrust fault at the base of the metamorphosed Franciscan exposed on the east shore. All three walks are past excellent exposures of schistose sandstone.

Description of the Island and Its Rocks

Angel Island is a single, sharp-crested mountain rising out of the flooded valley that is San Francisco Bay and separated from the mainland of Tiburon Peninsula by Raccoon Strait, part of the course of the Sacramento-San Joaquin River that flowed through the valley when sea level was 100 m lower than now and the shore was 45 km west of the Golden Gate. This phenomenon was caused by water being withdrawn from the oceans to form the great ice sheets of the ice age.

The plan of the island is roughly an equilateral triangle about 2 km on a side. Its summit, Mt. Caroline Livermore (MCL on Figure 36) is 238 m above sea level, and the average slope of the island exceeds 15° with cliffs along the shores. The summit is an east-trending ridge about 0.5 km long, from whose ends spurs descend to the most prominent points along the shore. The two that descend north to Point Ione and Campbell Point (see Figure 30)

enclose a sheltered valley and Ayala Cove, the main anchorage and ferry landing.

Two roads encircle the island: the Perimeter Road, mostly between 30 and 60 m above sea level, and the upper fire road, mostly between 120 and 150 m. Roads or trails descend to the shore at a number of points, but that to Blunt Point is closed to the public. A road reaches the summit from the south, and there are trails along several ridge crests. Without the excellent exposures along the roads and the shore, the geology of the island would be difficult to decipher, for only the serpentine has extensive natural exposures inland.

Shaded, north-facing slopes have a dense native woodland of live oaks, bay laurel, madrone, toyon, and buckeye. The south-facing slopes were probably orignally covered with grass or low chaparral but have been planted in part with eucalyptus and pines. Poison oak is a persistent shrub and discourages travel away from roads and trails.

For many years after the settlement of San Francisco Bay by Europeans in 1776, Angel Island was a watering place for sailing ships. In *Two Years Before the Mast*, Richard Henry Dana recounts two miserable wet days spent gathering wood on the island during the rainy December of 1835. From about 1863 until 1946 the island was in almost constant use as a military base, a military prison, an embarcation and debarcation center for three foreign wars, and an immigration and quarantine station. The buildings of Camp Reynolds (West Garrison) date back to the Civil War; those of Fort McDowell (East Garrison) to just before the World War I. The shore batteries on the southwest corner of the island were built during the Spanish-American War, 1898–1900. In the 1950's there were missile sites on the island. Parts of the island became a state park in 1954, and except for seven acres held by the Coast Guard at Blunt Point, the island has been a state park since 1966.

The Sandstone   The sandstone is medium to coarse, has locally thin layers of pebbles, and apparently is in turbidite beds (see Trip 3) 3 to 30 m thick that grade upward to silty layers. *Blake et al.* [1967] elsewhere recognized three zones of metamorphosed Franciscan sandstone, which they called textural zones I, II, and III. Sandstone of textural zone I shows no schistosity or other evidence of metamorphism visible to the naked eye. Sandstone of textural zone II has a well-developed schistosity, but with recognizable sand grains and with little or no evidence of recrystallization visible to the naked eye. On Angel Island, this schistosity is parallel to bedding in most places where both can be recognized. Textural zone III has been completely recrystallized to quartz-mica schist in which no sedimentary textures are preserved. Most of the sandstone of Angel

Island is in textural zone II, and the small bodies of schist are in textural zone III. The sandstone of Quarry Point is in textural zone I. *Schlocker* [1974, pp. 20–23] has an excellent description of the metamorphosed sandstone of Angel Island.

Microscopic study of the metamorphosed sandstone by Bloxam, Schlocker, and Coleman show it to be a typical greywacke made of grains of quartz, plagioclase feldspar, dark minerals, and volcanic rock fragments. Metamorphic minerals that have grown as microscopic crystals and crystal aggregates at the expense of these primary minerals and rock particles include jadeitic pyroxene, lawsonite, aragonite, and crossite. *Coleman* [1965] and *Schlocker* [1974] have shown that these new minerals grew without the addition of any material from outside the greywacke.

The Basalt   The basalt is in two bodies: a large body west of the serpentine dike and an arcuate belt within the sandstone east of the serpentine (Figure 31). The basalt west of the serpentine is submarine pillow basalt with thin interbeds of metamorphosed chert and shale (now schist) every few tens of meters. These were probably deposited on the basalt during intervals between eruptions. The interbeds generally strike northeasterly and dip 25°–35° northwest, suggesting that about 600 m of basalt is exposed.

The arcuate band of basalt east of the serpentine is about 300 m wide and has a discontinuous projection northward toward the southeast corner of Ayala Cove. The irregularities in its contacts convinced *Ransome* [1894] that it was an intrusion; however; in a few places there are suggestions of pillows and volcanic breccias [*Schlocker*, 1974, p. 35], indicating an origin by submarine eruptions. Its contacts with the sandstone are not generally exposed, but at the few exposures the contacts are zones of intense shearing, with much dark-grey to black clay gouge, suggesting fault movement. In most outcrops the basalt is more or less decomposed by weathering to a brown-to-orange mixture of iron oxides and clay minerals.

Microscopic study of the basalt by Bloxam and Schlocker showed that it originally consisted of pyroxene and plagioclase feldspar, but that in most of it the feldspar has been totally replaced by chlorite, lawsonite, and jadeite, and the pyroxene has been partly replaced by chlorite, glaucophane, and green jadeite. A dense mat of fine crystals of deep-blue amphibole (probably crossite) is also present.

The Serpentine   The serpentine dike is 50–175 m wide and appears to dip steeply west. The serpentine consists of ovoid masses of massive dense serpentine in intensely foliated material that resembles the slickentite of Baker's Beach (Trip 3). According to *Schlocker* [1974, p. 57], antigorite is the common serpentine mineral of the dense ovoid masses. Also present are ovoid masses and inclusions of gabbro, sandstone, and metamorphic rocks. Among the metamorphic rocks is rodingite, a greenish-grey rock consisting mainly of calcium silicate minerals such as hydrated grossularite garnet, vesuvianite, and diopside. Rodingite is thought to result from the alteration of basalt or gabbro inclusions within the serpentine [*Schlocker*, 1960]. The serpentine does not weather readily, hence it forms picturesque, craggy, greenish-grey outcrops in a belt across the west part of the island.

The Schist   The most remarkable rocks of Angel Island are the crystalline schists and other wholly crystalline metamorphic rocks. These are in small bodies a few meters thick and a few tens of meters long, within and on the borders of the basalt and serpentine. Their best exposures are along the shore, especially adjacent to the serpentine on the south shore and within basalt at Camp Reynolds. They appear to have been surrounded at the time they crystallized to schist by rocks that were not so extensively recrystallized.

In his microscopic study of the schist, *Schlocker* [1974, pp. 50–55] recognized many different kinds of schist, each consisting of a different assemblage of metamorphic minerals. These assemblages appear to depend on the original rock from which the schist was made (whether chert, sandstone, shale, basalt, or serpentine) and on the adjacent igneous rock (whether basalt or serpentine). Most of these assemblages can only be recognized through microscopic study of the rocks.

The Boat Trip

What To Take   Wear clothes you are willing to get dirty, wet, or scuffed. Boots with Vibram soles or tennis shoes are the best footgear. Bring warm sweaters, windbreakers, lunch, camera, hand lens, guidebook, and rain gear if it threatens to rain, all in a knapsack. *Do not* bring a hammer or geologic pick. Collecting rocks in the state park is forbidden, and outcrops of the most interesting rocks are limited and fragile. Their form and surface textures are a large part of their educational and esthetic value, and no amount of research information is justification for destroying these values for others.

When To Go   The best time to visit the outcrops along the shore is when the low tide falls between noon and 3 P.M.; but they can be seen anytime.

How To Get There   The ferry leaves from Fisherman's Wharf at the north end of Taylor Street (see Figure 23 for location) at 10 A.M. and 12 noon, Saturday and Sunday throughout the year and every day between June 1 and September 1. The last return boat leaves the island at 4:30 P.M. (somewhat later on some days in summer). It is illegal to miss the last boat, as there are no overnight accommodations, and camping is not allowed.

Municipal railway lines connect BART, the East-bay Terminal, and the Caltrans Depot with Fisherman's Wharf: the No. 42 (Downtown Loop), No. 19 (Polk), and No. 59 (Powell—Bay and Taylor cable car). Walk north on Taylor from any of these to the ferry dock.

The ferry usually passes close enough to Alcatraz Island to see the northeast-dipping beds of massive sandstone, in which Valenginian (Early Cretaceous—about 135 m.y.) fossils have recently been found [*Armstrong and Gallagher*, 1977].

If you round Angel Island on the east, you can see the offshore rocks of pillow basalt at Blunt Point—commonly occupied by seals—and the unmetamorphosed sandstone of Quarry Point along with the huge army barracks of East Garrison. The cliffs north of Quarry Point are of schistose sandstone. The immigration station was in the cove between Simpton and Campbell points.

If you round Angel Island on the west, you will be able to see cliffs of pillow basalt at Knox and Stuart points, flat-lying tan beds of Colma Formation behind Perle's Beach, and pale-green cliffs of serpentine and schist east of Perle's Beach. Camp Reynolds (West Garrison) lies in the valley between Knox and Stuart points, and the large building near the shore is a warehouse built of bricks brought around Cape Horn before 1877.

The ferry landing is at (1), Figures 36 and 37, where there are restrooms, drinking water, and a lunch room.

Walk A

The ultimate destination of Walk A is schist, serpentine, and pillow basalt at Perle's Beach (Figures 36–42). On the way it passes exposures of schistose sandstone and of serpentine on the west slope of Mt. Caroline Livermore.

Walk around Ayala Cove past the lunchroom to the greensward. The road cut between the lunchroom and the greensward exposes deeply weathered basalt. The building at the back of the greensward is the park headquarters and museum.

On the west side of the greensward, take a wooden path uphill past a restroom to a wide dirt path, and take this latter path to the right, first through a planted wood of acacias and Monterey cypress, then through a native wood of coast live oak, buckeye, laurel, and toyon. The path follows the ridge crest above Point Ione south to the Perimeter Road at 55-m (180-ft) altitude. Take a flight of steps on the far side of the road to the Mt. Carolyn Livermore trail, and follow the latter to the unpaved fire road at 130 m (425 ft) (2).

Here is a spectacular view of southern Marin County, the Golden Gate, and San Francisco. Directly west are the high, reddish hills of the Southern Marin Block (Trip 5; see Figure 16), with the scar of Highway 101 diagonally across their east face and Sausalito nestled at their northeast base. An abrupt change from steep slopes and brownish-red road cuts above and to the south to gentle slopes and grey road cuts below and to the north marks the thrust contact where the Southern Marin Block rests on the melange comprising most of the rolling hills of Marin County northward beyond Mt. Tamalpais. Mt. Tamalpais consists of "broken formation" and melange. It owes its height to resistance to erosion, provided by a tourmaline cement in the rocks of its summit. The tourmaline was probably deposited by boron-rich hot-spring waters before erosion had lowered the surrounding landscape below the summit of the mountain. Greyish-green exposures at the west end of the summit ridge of Mt. Tamalpais are serpentine. For other points of interest in Marin County to be seen from here, see Trip 2 and Figure 16.

The route from (2) is south along the fire road, past excellent road cut exposures of schistose sandstone—here a very coarse, dark, feldspar-rich greywacke. Layers of conglomerate in the sandstone are approximately parallel to the schistosity. Pebbles in the conglomerate are predominantly quartzite and grey chert, with rare pebbles of schist and altered volcanic rocks. Some pebbles are stretched, but most preserve their rounded form. A faint lineation on schistosity surfaces plunges down the dip of the schistosity.

Discontinuous sheets of dark-grey siliceous material 2–8 cm thick, and parallel to bedding in the sandstone, are present in road cuts around the next point and are identified by *Schlocker* [1974, p. 23] as siliceous shale and quartz veins.

At (3) the fire road crosses into basalt. The basalt is orange-brown where deeply weathered but is pale olive-green in deeper road cuts. It is intensely fractured but does not exhibit any schistosity. A road cut 30 m south of the canyon bottom has suggestions of pillow structure convex to the northeast, implying that the basalt is right side up. About 30 m farther the basalt contains thin veins of blue amphibole. The southern contact of basalt with sandstone is in the gully immediately south. Schistose sandstone is exposed in road cuts around the next spur.

At (4), take the trail downhill southwest along the ridge crest. The trail crosses red and green chert and onto serpentine, whose gnarled and craggy outcrops, draped with oaks and laurels, decorate the hillside to the northwest. The trail crosses onto basalt just northeast of a round water tank. Bands of fine silvery phyllite or schist in the basalt cross the trail.

The trail curves southeast (left) where the earthworks of Battery Wallace, one of the three Spanish-American War shore batteries, are hidden in the trees. Shortly beyond, the trail joins the Perimeter Road. At (5), about 245 m east of this junction, is a quarry in serpentine, mined for road metal and for a dam in the canyon to the east. The structure of the serpentine is well exposed in the quarry walls. Storage bins for the "crushed rock" stand east of the sharp bend in the road.

Return west along the Perimeter Road to the viewpoint above Knox Point and take the path on the left downhill through a eucalyptus forest to Perle's Beach. The route is over Colma Formation, a late Pleistocene deposit of dune sand, alluvium, and colluvium deposited during a glacial-age dry-valley stage of San Francisco Bay. Exposures of the Colma Formation in the beach bluff consist of horizontally bedded alluvium containing angular fragments of schist, chert, sandstone, and basalt, possibly derived from the valley east of (5), which is interbedded with brown sand and silt that was probably windborne from the sandy flats of the Sacramento-San Joaquin River, which flowed just west of here during the low sea-level stands. Intermittent erosion during the accumulation of the Colma is indicated by shallow unconformities exposed in the bluff.

Forty-five meters west of the foot of the access trail the Colma rests on weathered pillow basalt, which makes up the cliffs to the west. About 30 m west of this point, a layer of chert and schist about 3 m thick and dipping 40° northwest interrupts the pillow basalt and probably marks an interval between two submarine eruptions. The best exposure of pillows, showing that younger rocks are to the northwest, is at the extreme west end of the shore accessible from the beach and can only be reached at low tide (6; Figure 38).

About 30 m east of the foot of the access trail the Colma Formation is in contact with serpentine, and 25 m farther east a promontory of serpentine juts toward the Golden Gate (7, Figure 39). Foliation in the serpentine is irregular but dips, on the average, 30°–50° northwest and wraps around lenticular-to-ovoid bodies 2–20 cm across; most of these are massive serpentine, but a few are of other rocks. Eastward the serpentine is more sheared, is altered to white material (probably magnesite and hydromagnesite), and contains large tectonic blocks of sandstone. Nearly horizontal shear planes 2–6 m apart cut the foliation.

At the east end of the beach the serpentine is in

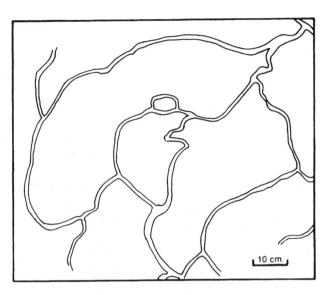

Fig. 38. Cross section of deformed pillows at locality 6, Figures 36 and 37, at the west end of Perle's Beach, indicating younger rocks to the west; view north. (Sketched from a photograph.)

contact with schist. Brown stilpnomelane-bearing schist with a texture almost of weathered wood alternates, in layers a few centimeters thick, with grey to dark-blue, massive, quartz-rich rock colored by glaucophane, crossite, and garnet. The parent material of the latter rock was probably chert and of the former shale. The original chert and shale beds were tightly folded, and the axial planes of the folds are parallel to the schistosity. The metachert layers are locally boudinaged (8, Figure 39; see Figure 40). Subsequently, the schist was compressed in a direction different from that which caused the schistosity, and kinks developed in the schistosity (Figure 41); these ultimately grew to intricate folds, deforming the schistosity (Figure 42). The contact of this schist with serpentine runs along the bluff about 10 m above sea level.

The glaucophane- and stilpnomelane-bearing schists are in fault contact on the east with a pale-green less-schistose crystalline rock, according to J. Schlocker (written communication, 1981), probably metamorphosed basalt. This rock makes a cliffed promontory that blocks the view and nearly blocks foot travel along the shore. The traverse around this point requires some agility as well as nonskid footwear. On the east side of the point, thin, discontinuous beds and pods of blueschist in this green rock strike northwest and dip steeply northeast, or are vertical. A large landslide has brought blocks of complexly folded yellow- to grey-weathering schist down to the shore in the cove east of this point (see Figure 36). Travel beyond (9) is not recommended, as the only access to the shore farther east is via a narrow deer trail etched into unstable colluvium

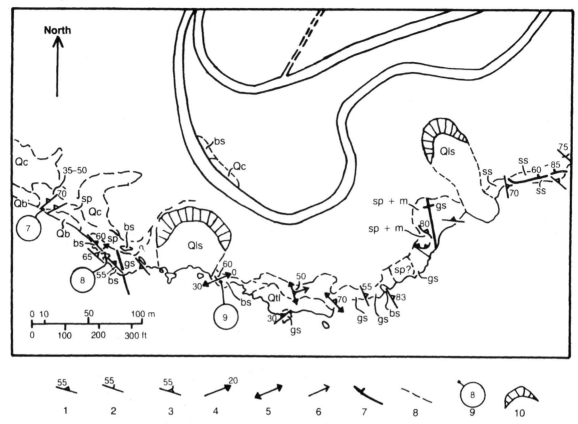

Fig. 39. Map showing the geology along the south shore of Angel Island, east of Perle's Beach (see Figures 36 and 37); based on original survey; base after 1:2400 enlargement of 1965 Caltrans BATS aerial photograph: **1,** strike and dip of schistosity; **2,** strike and dip of bedding; **3,** strike and dip of parallel bedding and schistosity; **4,** bearing and plunge of fold hinge line; **5,** bearing of horizontal fold hinge line; **6,** bearing and plunge of linear streaking in the plane of the schistosity; **7,** fault showing dip; **8,** contact and edge of exposure; **9,** numbered locality mentioned in text; **10,** headwall escarpment of landslide; **Qb,** modern beach sand; **Qtl,** talus; **Qls,** landslide deposit; **Qc,** Colma Formation; **bs,** blueschist; **gs,** green- to grey-colored schist; **sp,** serpentine; **m,** melange; **sp + m,** serpentine plus melange; **ss,** sandstone with interbedded siltstone (textural zone II).

perched perilously above vertical cliffs that plunge directly into the bay.

Retrace your steps to the Perimeter Road at Knox Point, and follow the Perimeter Road north. If time permits, you can combine walk A with part of walk B by taking the left-hand branch of the road, just north of Knox Point, which takes you to Camp Reynolds, from which you can get to the seawall along the bay shore; (10), at the south end of the seawall (see also Figure 37), is the beginning of exposures described in Walk B.

If time does not permit, take the right-hand fork, beyond Knox Point, which follows the contour around the head of the valley of Camp Reynolds. Exposures are very poor; however, craggy outcrops of grey-green serpentine can be seen on the hillside to the east. The exposures on the road north of (11), where it crosses from serpentine onto sandstone, are described in Walk B.

Walk B

Walk B is to exposures of chert and shale, metamorphosed to blueschist, in the pillow basalt at the west end of the island. It follows Walk A as far as (12), then follows the Perimeter Road south to Camp Reynolds, goes west through Camp Reynolds to the shore, and along the shore to the south end of the cove.

Schistose sandstone, locally pebbly, is well exposed in the road cuts between (12) and (11). For 245 m south of (12), schistosity (and apparently bedding) strikes north, parallel to the road, and dips 30°–40° west; southward, the schistosity and bedding bend abruptly, possibly at the axis of the major synclinal fold, to strike west across the road and dip steeply north. At (13), where a rockslide has taken place along the schistosity, flattened pebbles of black chert and grey-to-white quartz are in the sandstone.

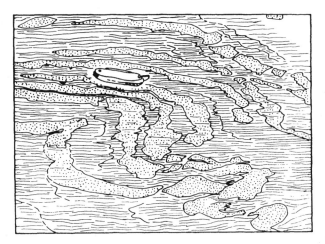

Fig. 40. Boudinaged metachert (stippled)—blue-grey quartz-crossite rock—interbedded with stilpnomelane-bearing schist; schistosity is parallel to axial planes of asymmetric folds in the original bedding. West end of wave-cut bench at locality 8, Figure 39. Schistosity trends N35°W. Top of drawing is to northeast. Knife is 9 cm long. (Traced from a photograph.)

South of the fold hinge, thin veinlets of white quartz appear in the sandstone. At (14) the road goes through a large cut, and the rock is foliated enough to be called a schist. At the south end of this cut, a path leads to the right (west) to a bench and picnic table on an abandoned roadbed, where irregularly distributed, extremely fine-grained, pale-blue schist is exposed in basalt. Basalt is exposed in the next outcrop to the south on the Perimeter Road. The swale south of this basalt has tules and willows, indicating water close to the surface, which is probably the result of impervious clay along the contact of the basalt with the sandstone exposed in

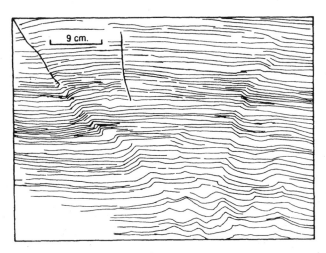

Fig. 41. Kink bands in schist at back of wave-cut bench at locality 8, Figure 39. Schistosity strikes N45°W, dips 67°NE. Kink on left side strikes due north, dips vertical; kink on right side strikes N75°E, dips 35°N. (Traced from a photograph.)

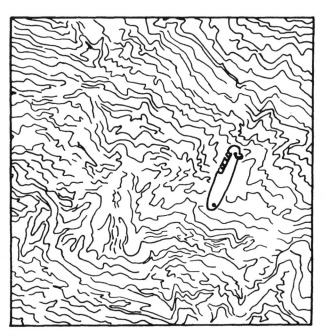

Fig. 42. Complexly folded schistosity. Traced from a photograph of part of a block of schist in a landslide deposit in a cove halfway between localities 8 and 9, Figure 39. View is along hinge lines of folds in schistosity. Knife is 9 cm long.

the high road cut just south of there. This sandstone is cut by numerous quartz veins. A large black schist boulder in the east bank of the road, about 10 m north of the fire hydrant at (11), marks the contact between sandstone and serpentine.

Follow the road downhill to Camp Reynolds, and follow the path alongside the officer's quarters on the south side of the parade ground to the seawall. This was probably constructed between 1864 and 1877 of sandstone from Quarry Point, with an occasional gabbro or granite boulder of unknown origin. The sandstone blocks of the lower half of the seawall, which are subject to wave attack at high tide, have not been noticeably eroded. But the blocks in the upper half of the seawall, subject only to salt spray, have been etched into a complicated honeycomb pattern and have retreated an average of 15 cm from the surface defined by the basal blocks and by the occasional blocks of gabbro and granite in the upper part of the wall. This rapid weathering is probably the result of salt-crystal growth during evaporation of the spray; the basal blocks may never dry sufficiently to precipitate the salt crystals that pry the sandstone apart.

From the south end of the seawall, walk south along the beach. A sketch map of the south end of the cove is shown in Figure 43. The first exposure encountered is pillow basalt with thin seams of pale-blue schist. About 30 m south the pillow basalt rests on a breccia of basalt fragments that are 0.5–10 cm

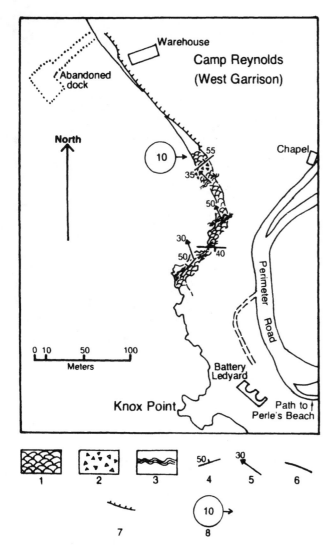

Fig. 43. Sketch map of the geology of the south end of the cove at Camp Reynolds (West Garrison) (see Figures 36 and 37 for location), based on original survey; base after 1:2400 enlargement of 1965 Caltrans BATS aerial photograph: **1,** pillow basalt; **2,** volcanic breccia; **3,** schist with metachert; **4,** strike and dip of sedimentary bedding; **5,** bearing and plunge of fold hinge line in schist and metachert; **6,** fault; **7,** sea wall; **8,** numbered locality mentioned in text.

across; the crude layers of breccia dip 55°NW; coarsely crystalline calcite fills interstices and veins the breccia. The breccia rests on more pillow basalt, which has layers of tightly folded blueschist every 10–30 m along the shore to the south. Hinges of the folds plunge 40°–50°NW. The schist is cut by veinlets of quartz, calcite, and zeolites (?). Pillow tops indicate that the section is right side up. The southernmost exposure of the schist has massive blue layers 2–8 cm thick that have been boudinaged and which are apparently layers of chert that has been metamorphosed to a rock consisting of quartz and blue amphibole.

The cliff at the north end of the seawall also ex-

poses sheared and altered pillow basalt; the shore platform at the base of this cliff is accessible at low tide and shows irregular layers of blueschist similar to the blueschist on the south side of the cove. After a picnic or a sunbath, you can retrace your steps to the ferry landing or, if time permits, return along the upper fire road via part of Walk A.

Walk C

Walk C is to the flat fault zone and underlying unmetamorphosed sandstone at Quarry Point on the east side of the island. At the ferry landing, go north at the restrooms, through the picket fence, past the base of the red chert exposure, and up the hill, through a small picnic area on a shelf 15 m above the bay, then on a flight of steps to the Perimeter Road about 50 m above the bay. Follow the Perimeter Road left (east) past medium-fine schistose sandstone with rare shale beds. At (15), on the east side of the first canyon, the sandstone sequence is in a broad anticline. The sandstone south of the anticline is thin-bedded, with shale and some 1–2 cm bands of grey siliceous material that may be chert. Where the road bends around the ridge south of Campbell Point, schistosity and bedding abruptly change direction to strike northwest and dip steeply east; no sign of the anticline can be found in road cuts on the east side of the ridge. Such abrupt changes in direction of bedding and schistosity are common on Angel Island.

Road cuts expose east-dipping sandstone as far south as the point overlooking the cove of the quarantine station; beyond this point exposures are poor. The next good exposure is at (16) and is a quarry in basalt so deeply weathered as to be almost unrecognizable. The road continues in basalt around the ridge from Simpton Point. The road that goes uphill past the firehouse and park employee dwellings leads to better exposures of basalt. At (17), in the bank of an abandoned road that can be reached only from the upper fire road, is pale-green basalt cut by curved fractures lined with silky blue amphiboles.

The route to Quarry Point is south along the main perimeter road. About halfway between the fire station and the left turnoff to Quarry Point the road crosses the buried contact between basalt and sandstone. Schistosity in the sandstone dips gently southwest. Take the left (east) fork for Quarry Beach, past massive barracks buildings of East Garrison. Keep to the northernmost road until you have a clear view of the shore to the north. At low tides it is possible to clamber down to the beach and walk north along the shore to beyond Simpton Point. The quarry itself is fenced off, but it is possible to examine the sandstone of Quarry Point at the north end of the quarry wall (18) that is just outside the fence. Its difference from the sandstone of the rest of the island, and its similarity to the sandstone of San Francisco, is immediately apparent.

For about 215 m north of the end of the artificial seawall (19) the wave cut platform exposed at low tide in winter is cut on what appears to be the unmetamorphosed sandstone of Quarry Point. Resting on this platform are scattered blocks, from 1 to 4 m in maximum dimension, of the following rocks: diabase, chert cobble conglomerate, folded chert, yellow and red jasper, light-grey silicified sandstone, and pillow basalt containing closely packed light-colored spherical masses in a darker matrix (These are called varioles and result from the growth of spherical masses of crystals when the originally glassy basalt devitrified). This is a typical assemblage of exotic blocks such as is found in melange (see Trip 4, Side Trip B). The south end of the bluff behind the beach is densely overgrown with willows, indicating groundwater close to the surface and many springs, which in turn implies the outcropping edge of a body of impermeable clay fault gouge. Farther north are extensive landslides and earthflows, whose toes expose black shale and clay. At (20), where exposures along the base of the bluff first appear to be in place, the planes of shearing in the black clay-rich material of the bluff are nearly flat, implying a horizontal fault or flat body of melange.

The bluff steepens northward, and outcrops of typical schistose sandstone are 5–10 m above sea level. The base of this sandstone descends to sea level a few meters to the north. The shore northward beyond Simpton Point exposes only the tan-weathering schistose sandstone and is accessible only at low tide.

This sandstone is strongly schistose, locally conglomeratic, locally has interbeds of siltstone and shale, and seems to consist of massive graded beds—the usual product of turbidite sedimentation. In general, the bedding and schistosity strike parallel to the shore and dip gently west toward the island but are locally folded and faulted. Pebbles and cobbles are quartz, grey chert, and pale-green to white volcanic and granitic(?) rocks, and they are well rounded but have been tectonically flattened and locally boudinaged. Direction of maximum stretching appears to be southwest down the dip of bedding and schistosity.

At (21), about 250 m northwest of Simpton Point, the geology is quite complex, with a layer of thoroughly crushed shale resembling melange matrix separating the sandstone from a narrow band of basalt on the west. Near the crushed shale the sandstone is intensely folded. East of the zone of tight folding, thin shale partings in the sandstone have abundant coalified plant remains, much like those at

the north end of Baker's Beach (Trip 3, locality 1, Figure 15). A point of sandstone west of the basalt prevents foot passage into the cove of the quarantine station.

On the south side of Quarry Point is a quiet, sheltered, sandy beach that is an excellent place for a sunbath, swim, or picnic. Behind the beach is a narrow belt of dunes, and behind the dunes is a bluff of sandy colluvium with a thick soil, part of the Colma Formation and perhaps the original source of the sand on the beach. Quarry Point sandstone is exposed at the northeast end of the beach. At (22), at the southwest end, is an exposure of apparently nonschistose sandstone in graded beds 3–5 m thick (with 0.5–1 m shale interbeds) that dip 50° southwest and are possibly overturned. The shale is sheared and contains tectonically rounded blocks of sandstone. Loose blocks on the beach and shore platform at the base of the sea cliff here include serpentine and sheared conglomerate and were derived from a small landslide. They may have come from a melangelike fault zone hidden in the willows above, which would be the southward continuation of the fault exposed north of Quarry Point.

The return to the ferry landing can be made by retracing your steps or, if time permits, by following either the Perimeter Road or the upper fire road around the island. Whichever way you return, don't miss the last boat!

Trip 7. After Subduction is Over: A BART Trip to a Transform Fault

Introduction

Subduction ceased in San Francisco when the East Pacific Rise encountered the Cordilleran Subduction Zone. The Pacific Plate is now in contact with the North American Plate throughout most of California and is moving northwesterly past North America at a current rate of about 6 cm per year along the transform fault boundary known as the San Andreas Fault System. This is not a single fault break, although the San Andreas Fault dominates. Rather it is a zone 80–100 km wide in which there are several right-lateral strike-slip faults spaced a few kilometers to a few tens of kilometers apart. (A strike-slip fault is one along which the displacement or motion has been horizontal, i.e., parallel to the strike; if, as you face the fault, its far side has moved to your right, it is a right-lateral fault; if to your left, it is a left-lateral fault).

In the San Francisco Bay region the displacement

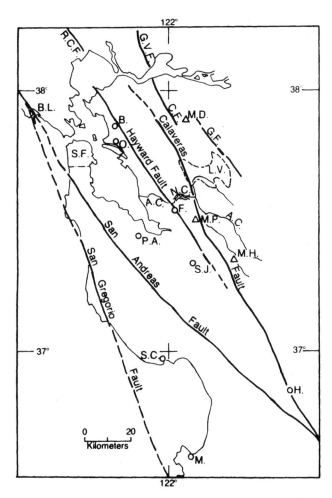

Fig. 44. Outline map showing major active right-lateral faults of the San Andreas system in the San Francisco Area; redrawn from *Jennings* [1977]: **A.C.**, Alameda Creek; **B.**, Berkeley; **C.F.**, Concord Fault; **F.**, Fremont; **G.F.**, Greenville Fault; **G.V.F.**, Green Valley Fault; **H.**, Hollister; **L.V.**, Livermore Valley; **M.**, Monterey; **M.D.**, Mt. Diablo; **M.H.**, Mt. Hamilton; **M.P.**, Mission Peak; **N.C.**, Niles Canyon; **O.**, Oakland; **P.A.**, Palo Alto; **R.C.F.**, Rodgers Creek Fault; **S.C.**, Santa Cruz; **S.F.**, San Francisco; **S.J.**, San Jose.

appears to be distributed among at least four major strike-slip faults (see Figure 44): the San Gregorio Fault, the San Andreas Fault, the Hayward Fault, and the Calaveras Fault. The Hayward is the most accessible of these faults. Good views can be had of the topographic evidence of geologically recent fault movement (Figure 47) from the BART trains and stations on the Fremont Line. Within a short walk from the Hayward BART station, it is possible to see curbings and other man-made features offset by creep along the fault. Major earthquakes occurred on the fault in 1836 and 1868.

For most of its length, the Hayward Fault runs along the boundary between the San Francisco Bay plain and the ridge of hills to the east. At its north end this ridge is known as the Berkeley Hills, and between Hayward and Fremont it is called Walpert

Ridge. In places, vertical displacement along the fault has uplifted the hills, but in the north the fault is within the hills and even along ridge crests, and in the south it leaves the hills to cross the alluvial plain.

The Rocks East of the Hayward Fault

The Hayward Fault is also the boundary between the Franciscan and other rock sequences to the east—important to plate tectonics in California. These rock sequences can be seen in the Berkeley Hills and along the fault from the BART train. There are two groups, the Great Valley Sequence of Jurassic and Cretaceous age—contemporaneous with the Franciscan—and a much younger Miocene and Pliocene sequence of marine and land-laid sedimentary rocks interbedded with lavas and tuffs. Both sequences are tightly folded, and their beds dip steeply and in places are even overturned [*Hall*, 1958; *Robinson*, 1956; *Radbruch*, 1969].

The Great Valley Sequence is a turbidite sequence of sandstone, shale, and conglomerate that in Mesozoic time accumulated in a deep ocean trench that lay between the subduction zone and the land area of North America [*Ingersoll*, 1979]. It originated in a manner similar to that of the sandstone and shale of the Franciscan, except that it was deposited on a seafloor that was above the subduction zone and was therefore never subducted (Figure 45*b*). In consequence, individual beds can be traced long distances, and there is no broken formation or melange.

In a few places in the Coast Ranges, the lowest beds of the Great Valley Sequence rest on submarine lavas, diabase dikes, and gabbro, interpreted to be the ophiolite (i.e., oceanic crust) on which the Great Valley Sequence was deposited [*Page*, 1981; *Hopson et al.*, 1981]. However, throughout most of the Coast Ranges, sedimentary rocks of the Great Valley Sequence are in contact with the Franciscan along a great fault called the Coast Range Thrust, which is thought to mark the subduction zone along which the Franciscan was subducted (Figure 45; *Blake and Jones*, [1981]). Great sheets of ultramafic rock (now largely altered to serpentine) are present along this fault and are thought to have been squeezed up along the thrust from the mantle. The Coast Range Thrust has been folded and faulted so that down-folded or down-faulted patches of the Great Valley Sequence, in fault contact with the Franciscan, are found throughout the Coast Ranges west of the main exposure of the Coast Range Thrust, which is on the east side of the Coast Ranges. The Great Valley Sequence along the Hayward Fault is one such patch (Figure 45*a*).

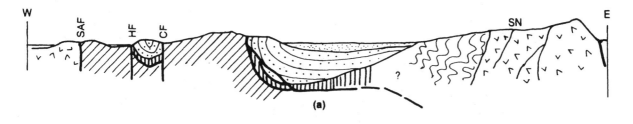

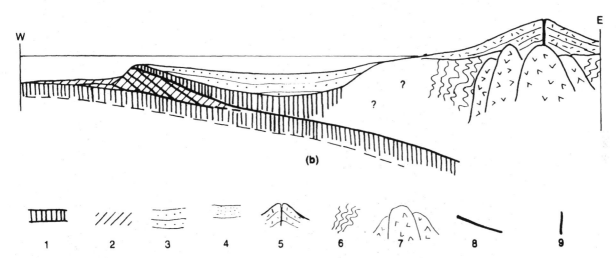

Fig. 45. Diagrammatic cross sections through the central Coast Ranges and Sierra Nevada, showing relationships of the Great Valley Sequence to the Franciscan Formation: (a) diagrammatic, present-day relationships along an approximate east-west line through Fremont, not to scale; (b) Hypothetical relationships at the time of accumulation of the Great Valley Sequence (adapted from *Ingersoll* [1979, Figure 8]): **1,** oceanic crust or ophiolite sequence; **2,** Franciscan rocks (structural details not shown in section (a)); **3,** Great Valley Sequence; **4,** Cenozoic rocks of the Great Valley; **5,** Cretaceous volcanos of the Sierra Nevada; **6,** pre-Cretaceous metamorphic rocks of the Sierra Nevada; **7,** Mesozoic plutonic rocks of the Sierra Nevada; **8,** thrust fault; **9,** vertical fault (**SAF,** San Andreas Fault; **HF,** Hayward Fault; **CF,** Calaveras Fault); **SN,** Sierra Nevada.

Between the active strand of the Hayward Fault and various faults in a 1.5–2 km belt to the east that are thought to be formerly active strands of the fault are extensive bodies of gabbro that may have been broken from the ophiolite beneath the Great Valley Sequence and squeezed up along the fault; there are also bodies of serpentine that could have been squeezed up along the Hayward Fault from the ultramafic rock along the Coast Range Thrust.

The Miocene and Pliocene rocks are found in the northern Berkeley Hills and the hills east of Fremont, as well as in a large area northwest of Livermore Valley (Figure 44). They include Miocene diatomaceous shale and chert and late Miocene and Pliocene conglomerate, sandstone, and mudstone, which were deposited on a great alluvial plain. Basalt flows and rhyolite ash falls that erupted about 10 m.y. ago from centers in the Berkeley Hills are interbedded with the land-laid sedimentary rocks. The tight folding, which has in places overturned

these rocks, may be a secondary effect of movement on the Hayward Fault.

The BART Trip

Board any Fremont train at any BART station in San Francisco (Figure 46). On Sundays, take the Concord train to 12th Street Station in Oakland and transfer to the Fremont train. If you plan to take the walking trip at Hayward, purchase a ticket equal to two one-way trips between your origin and Hayward; you use the remainder of the ticket to complete your journey after the walking trip. If you do not plan to take the walking trip, you will not have to leave the BART-paid area until you return, and you should consult the station attendant about the purchase of a BART excursion ticket.

The train goes under the bay in a double tube 6 km long, reaching a maximum depth of 41.5 m be-

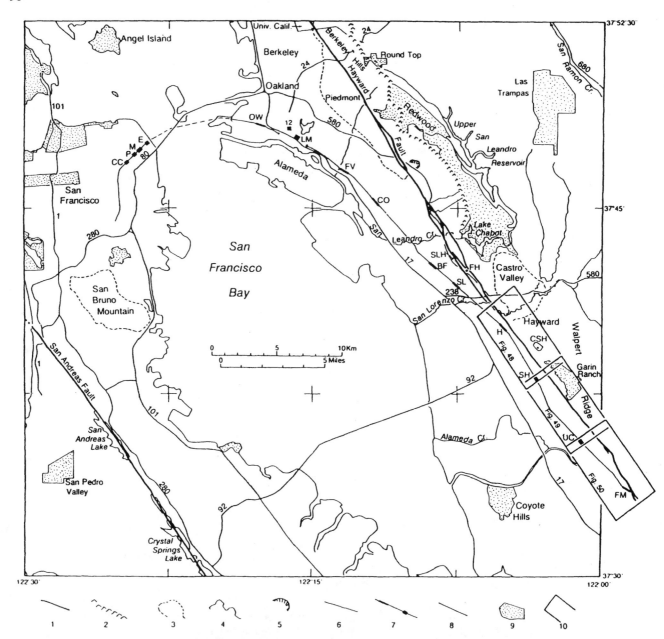

Fig. 46. Outline map of part of San Francisco Bay Area, showing the BART Fremont Line, the San Andreas and Hayward faults, freeways, regional parks, and some localities mentioned in Trip 7 (redrawn from U. S. Geological Survey 1:100,000 map of San Francisco 1° x 30′ sheet, 1978): **1,** Hayward and San Andreas faults; **2,** approximate crest of Berkeley Hills; **3,** approximate base of hills in Berkeley, Oakland, around Castro Valley, and around San Bruno Mountain; **4,** shore of bay, lake; or stream; **5,** quarry; **6,** BART line; **7,** BART station—see list at the end of explanation; **8,** freeway; **9,** park; **10,** location of Figures 48, 49, and 50; **SLH,** San Leandro Hospital; **FH,** Fairmont Hospital; **CSH,** California State University at Hayward; BART stations: **CC,** Civic Center; **P,** Powell; **M,** Montgomery; **E,** Embarcadero; **OW,** Oakland West; **12,** Twelfth Street; **LM,** Lake Merritt; **FV,** Fruitvale; **CO,** Coliseum; **BF,** Bay Fair; **SL,** San Leandro; **H,** Hayward; **SH,** South Hayward; **UC,** Union City; **FM,** Fremont.

low sea level. It is above ground and elevated for a short distance in west Oakland, where you can see the Berkeley Hills, their highest points being the outcropping edges of late Miocene lava flows. The train is underground again through central Oakland. Beyond the Fruitvale Station, features related to the Hayward Fault come into view on the left

(NE) side of the train as it proceeds toward Fremont (Figure 46).

In front of the Berkeley Hills is a low ridge covered with houses, behind which the hills form the skyline. The low ridge is of Franciscan, and the skyline ridge of Great Valley and Tertiary rocks. The Hayward Fault runs in a narrow valley between the

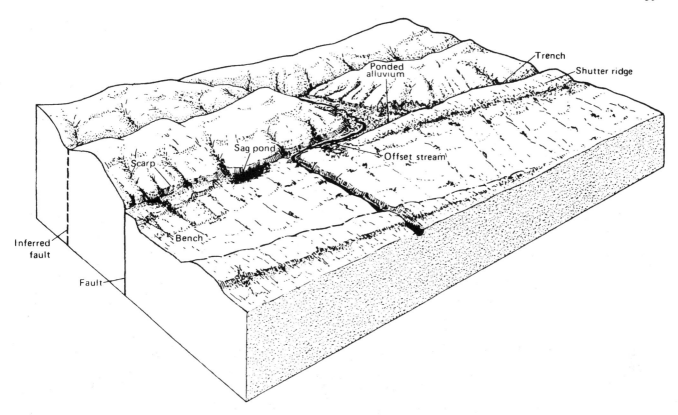

Fig. 47. Diagram showing some of the topographic features resulting from faulting [from *Radbruch-Hall*, 1974].

two ridges. A large light-brown area on the face of the ridge east of the fault is a huge quarry for crushed rock in the Leona Rhyolite, a late Pliocene or early Pleistocene pyrite-bearing, sodium-rich rhyolite that was erupted along the Hayward Fault as a series of domes [*Robinson*, 1956; *Radbruch*, 1969]. Halfway between the Coliseum and San Leandro stations the near ridge dies out, and southward the Hayward Fault lies along the southwest base of the Berkeley Hills.

At San Leandro Station you can see the canyon of San Leandro Creek, which drains Lake Chabot and San Leandro reservoirs and a large part of the Berkeley Hills. The greyish-green rock visible in road cuts in the hills southeast of San Leandro Creek is gabbro. The active strand of the Hayward Fault runs along the bench behind the lowest hills. Several small domes of Leona Rhyolite cap the hilltops [*Robinson*, 1956]. The large building against the base of the hills is the San Leandro Hospital, about 300 m west of the active strand of the fault.

The cluster of red-roofed buildings east of the freeway due north of Bayfair station is Fairmount (Alameda County General) Hospital; the active trace passes beneath a couple of the buildings. South of Bayfair Station the train descends to ground level, passes beneath Highway 238, and rises again. As seen from the train between Bayfair Station and Highway 238, the fault runs along the freeway (Highway 580) at the base of the hills.

Beyond Highway 238, BART crosses the concrete-lined channel of San Lorenzo Creek, which drains Castro Valley, a shallow semicircular downwarp about 4 km across, within the hills. The creek emerges from Castro Valley about 0.8 km north of the Hayward BART station (Figures 46 and 48). It bends sharply to the right to flow for about 1.5 km northwest along the fault before turning southwest toward San Francisco Bay. About 1 km of this abrupt jog is due to recent uplift of a narrow bedrock ridge along the fault, but as much as 0.5 km could be the result of right-lateral displacement along the fault.

Hills of gabbro and Leona Rhyolite are around the Hayward City Office Building, the tall striped structure in the distance on the left. As the train approaches Hayward Station, the narrow bedrock ridge blocking San Leandro Creek can be seen a few blocks northeast of the tracks (Figure 48).

Hayward BART station is the point of departure of the walking trip, which can be made either before or after the train trip to Fremont (see below). The gap in the hills opposite the station is the mouth of Castro Valley. The hilltops rise gradually south of this gap, and the hills face the bay plain with a straight and locally benched escarpment that

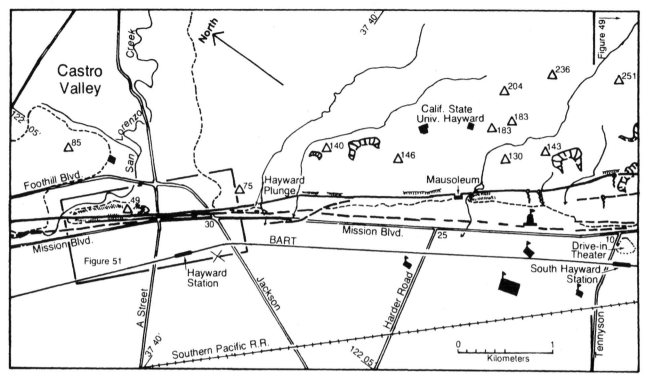

Fig. 48. Sketch map showing BART line, Hayward Fault, and topographic features between Hayward and South Hayward BART stations; traced from *Radbruch-Hall* [1974]; see legend at Figure 50.

is probably the result of uplift along the fault. The scarp is breached every kilometer or so by narrow canyons that drain the interior of the hills; many of these are offset by the fault, which runs along the lower slope or at the base of the escarpment. Topographic features that demonstrate the presence of the fault are visible from the BART train between the Hayward and Fremont stations. Many of these are illustrated in Figure 47.

Beyond Jackson Street (Figure 48), BART is at ground level, and trees and buildings hide the hills. When we again see the hills, there is a bench about a third of the way to their top. The bench is covered with houses, and the scarp in front of the bench is as straight, steep, and smooth as a road cut. The active trace of the fault is along the base of this smooth, straight scarp, whose unusual character is probably due to vertical uplift along the fault.

The tall square tower on the skyline is the administration building of California State University at Hayward, whose campus is out of sight on flat hilltops. At the foot of the hills below the tower is a pink, churchlike building—a mausoleum—with a cemetery we can barely glimpse before a supermarket blocks the view. This mausoleum is beside a canyon whose mouth is blocked and diverted northwest by a ridge that has been moved across it by the fault (Figure 48).

When the hills come in view again, the building

seen at their base is a parochial school. Behind the school, many of the small streams and ravines coming down the slope have a sharp kink to the northwest (that is, to their right, your left) as they cross the active trace of the fault, about a third of the way up the hillside. These kinks are also the result of fault offset. The active strand of the fault just beyond is about 60 m above the plain, behind rows of identical grey-roofed white-walled apartment houses on the lower bench. The high skyline hills behind the large quarry northeast of the South Hayward station are part of the Garin Ranch Regional park (Figure 46), a new wilderness park of the Eastbay Regional Park system and within walking distance of BART.

At South Hayward, BART climbs back to elevated status, crosses a drainage canal, and passes a golf course (Figure 49). Ravines coming straight down the mountain front beyond the golf course are all kinked sharply to the north, the best example in the Bay Area of streams offset by fault movement. Past the golf course, BART crosses high over the Southern Pacific Railroad and descends into a shallow trench, and the view is partially blocked by nearby houses and trees. On a bench about 60 m above the plain is a large, pale-green water tank, which is athwart or very close to the active trace of the fault [*Radbruch-Hall*, 1974].

The large pink building just above the foot of the

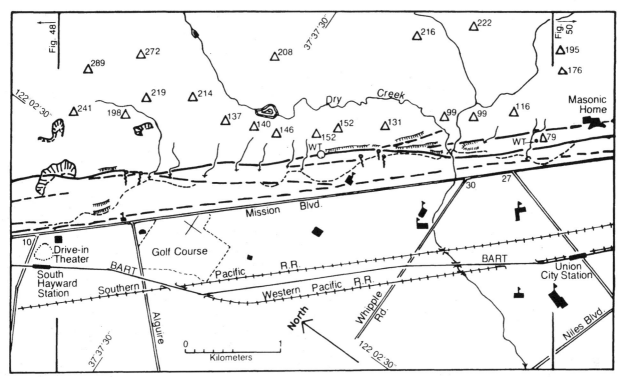

Fig. 49. Sketch map showing BART line, Hayward Fault, and topographic features between South Hayward and Union City BART stations; traced from *Radbruch-Hall* [1974]; see legend at Figure 50.

hills directly opposite Union City BART station (Figures 49 and 50) is the Masonic Old People's Home. The most recently active strand of the fault, along which there was movement in the 1868 earthquake, lies at the foot of the slope in front of the Masonic Home. An older, probably inactive, strand passes through the small green tank on the hill northwest of the home. Southeast of the home, the active strand lies at the foot of the low escarpment in front of a housing tract on a bench and is marked by groves of willows at springs and seeps.

Nearby on the left, beyond the Union City Station, are steel mills, and on the right in the distance are the Coyote Hills, an island of Franciscan rocks in a sea of alluvium and bay mud. The hills are bordered by salt-evaporating ponds and salt marshes rich in bird life and which are now a regional park (Figure 46). Southeast of the Niles Boulevard overpass, large gravel pits are on the right (SW). Their banks expose gravel of the alluvial fan of Alameda Creek, which drains Livermore Valley and a large area of the Diablo Range. Alameda Creek crosses the hills west of Livermore Valley through Niles Canyon (see Figure 44), a narrow meandering gorge that appears to have been cut by the creek as the hills were slowly uplifted across its course in Pleistocene time. The gravel is derived largely from the Diablo Range. This alluvial fan is an important groundwater reservoir, and low dams in the bed of

Alameda Creek—crossed next by BART—keep creek water seeping into the gravel to replenish this reservoir. The gravel mined from these pits, and from similar gravel pits in Livermore Valley, is the main source of concrete aggregate for much of the Bay Area.

Niles Canyon is visible to the east as the train crosses Alameda Creek and the Southern Pacific railroad tracks. The mountain front has bent abruptly east and makes an impressive scarp between Niles Canyon and Mission Peak, the high sharp-pointed peak to the southeast (Figure 44). The mountain front is a scarp along the probably active Mission Fault.

Where the mountain front bends abruptly east, the Hayward Fault continues in a nearly straight line across the alluvial plain and passes just east of the Fremont BART station (Figure 50). For a short distance across the plain, the west side of the fault has been lifted above the east side. The station is situation on the a low flat hill of alluvium pushed up on the west side of the fault. On the east side of the fault, clearly visible from the station platform, is a sag pond occupying a block of ground that has subsided between two strands of the fault. Gouge along the fault acts as a watertight seal, keeping groundwater east of the fault high and the pond filled with water.

Southward along the fault, behind some grey

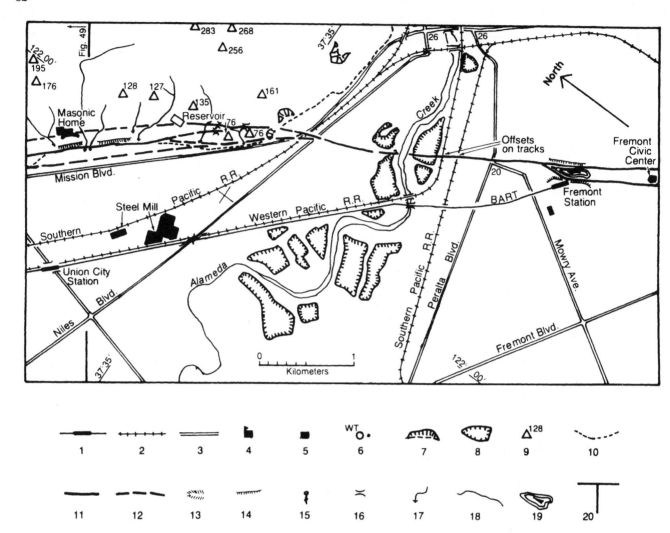

Fig. 50. Sketch map showing BART line, Hayward Fault, and topographic features between Union City and Fremont BART stations; traced from Radbruch-Hall [1974]. Legend: **1,** BART line with station; **2,** other railroad; **3,** major road; **4,** school; **5,** other landmark building; **6,** water tank; **7,** back well of quarry; **8,** gravel pit; **9,** hilltop, altitude in meters; **10,** edge of plain; **11,** active strand of Hayward Fault; **12,** possible inactive fault strand; **13,** shutter ridge; **14,** escarpment at front of bench; **15,** spring; **16,** saddle; **17,** offset ravine; **18,** stream; **19,** pond; **20,** boundary of adjacent Figures 48, 49, or 50.

buildings, is a low plateau that ends abruptly south-west against the fault. This plateau, located just east of the community of Irvington, consists entirely of poorly consolidated gravels that were uplifted and titled along the Hayward Fault. Remains of Pleistocene mammals, collected for many years from quarries in these gravels (now buried by a freeway), are the basis for the Irvingtonian land mammal age of North America, thought to encompass the time between 1.5 and 0.2 m.y. ago [*Berggren and Van Couvering*, 1974].

Without leaving the paid area of the station, you can take the next BART train north to the Hayward Station for the fault creep walk or the next San Francisco train to your starting point.

Walking Trip From the Hayward Station to See Evidence of Fault Creep

Walk south along the northeast side of the BART station, and turn left (NE) on C Street (Figure 51). Walk northeast along C to Mission Boulevard, cross Mission, and walk one block southeast (right) on Mission to D Street. In the block between Mission Boulevard and Main Street, between 1922 and 1971 the curbing on D Street was offset about 30 cm by creep along the Hayward Fault, the offset being concentrated along two active strands (Figures 51 and 52). If the street has not been recently repaired, en echelon cracks in the asphalt indicate where the active strands cross D Street. These cracks develop

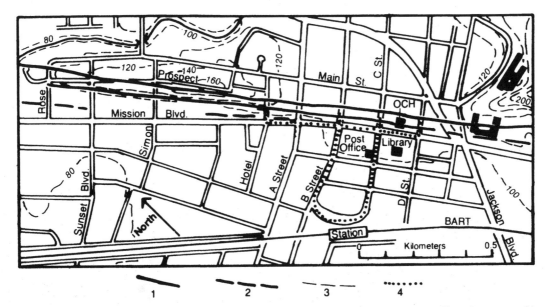

Fig. 51. Sketch map showing walking tour from Hayward BART station to offsets by creep along the Hayward Fault; traced from *Radbruch-Hall* [1974]; see Figure 48 for location: **1,** active strand of fault; **2,** possible inactive strand; **3,** topographic contour—contour interval 20 feet (6.1 m); **4,** route of walking tour; **OCH,** old court house.

as pavement is stretched by the lateral motion along the fault.

Return to C Street and examine the curbing on the north side of C between Mission and Main. An offset in the curbing can be seen in front of 934 C Street, and sighting along the wall of the building will show that one pillar is out of line.

One block north, on B Street, there is an abrupt rise just east of Mission Boulevard. The offset along the fault is concentrated at Nos. 921, 929, and 927, on the south side of B, and between Nos. 912 and 926 on the north side. Sidewalk squares are offset in front of 914. Curbing here is offset 5–8 cm. The offsets can be seen by sighting along building walls and curbing.

This part of town is the original settlement of Hayward, probably because groundwater was forced to the surface by clay gouge along the fault and provided springs and shallow wells.

One block north of A, along Mission Boulevard, is Hotel Avenue, which is at the south end of the 20-m-high and 900-m-long ridge of bedrock squeezed up along the fault that diverts San Lorenzo Creek to the northwest. The abrupt west wall of this ridge is a fault scarp. The active fault strand is part way up this scarp.

About half a block up Hotel Avenue, a frame house at No. 923 is on the active strand. The northwest corner of its foundations has moved some 20–25 cm out of line with the rest of the house, twisting the wall and front steps. A line of en echelon cracks extends across Hotel Avenue to the middle of the narrow street on the north.

The next streets crossing the fault, Simon and Sunset, some distance to the northwest along Mission Boulevard, show evidence similar to that on B, C, and D streets and Hotel Avenue. You can return to the BART station via B Street.

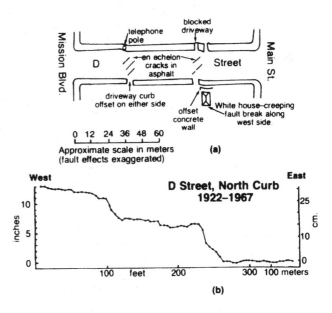

Fig. 52. (a) Sketch map, made November 28, 1976, of D Street between Main Street and Madison Boulevard, showing location of offset curbs and buildings. (b) Record of precisely surveyed creep displacement. The line with dots represents the position of the curb in 1967, which was placed perfectly straight along D Street in 1922 (from *Nason* [1971]).

Appendix A. Geologic Time Scale

Boundary in Million Years	Era	Period	Epoch	Age Shown Only for Jurassic and Cretaceous
0.01	Cenozoic	Quaternary	Holocene	
1.9			Pleistocene	
5		Tertiary	Pliocene	
22.5			Miocene	
37			Oliogocene	
55			Eocene	
65			Paleocene	
	Mesozoic	Cretaceous	Late (Upper[1])	Maestrichtian Campanian Santonian Coniacian Turonian Cenomanian
100			Early (Lower[1])	Albian Aptian Barremian Hauterivian Valanginian Berriasian
140		Jurassic	Late (Upper[1])	Tithonian Kimmeridgian Oxfordian
			Middle	Callovian Bathonian Bajocian Aalenian
176			Early (Lower[1])	Toarcian Pliensbachian Sinemurian Hettangian
195		Triassic		
230	Paleozoic	Permian		
280		Carboniferous	Pennsylvanian Mississippian	
345		Devonian		
395		Silurian		
435		Ordovician		
500		Cambrian		
570	Precambrian			
4000+				

Geologic time scale taken from *Van Eyesang* [1975].

[1]Applies to rocks deposited during the epoch.

Glossary

A

algae One-celled, colonial, or filamentous aquatic plants that lack a vascular system.

alluvial plain A plain formed by the deposition of alluvium.

alluvial fan A fan (land form) built by the deposition of alluvium (cf. **fan**).

alluvium A deposit of gravel, sand, or clay made by streams or mudflows.

ammonites Extinct cephalopod mollusks whose many-chambered, coiled shells had intricately convoluted sutures. Important guide fossils for the Mesozoic Era.

amphibole (mineral) A group of silicate minerals whose crystal structure involves double chains of silicon-oxygen tetrahedra. They form rods, needles, or fibers along which two cleavages intersect at 120°.

amplitude (of a fold) The distance, parallel to the axial plane, between the hinges on the same bedding surface of an adjacent syncline and the anticline (by analogy with waves.)

anastomosing Branching and rejoining irregularly to form a netlike pattern.

andesite Dark, fine-grained volcanic rock of intermediate composition (55%–65% silica), which may have phenocrysts of plagioclase and either augite or hornblende in a fine-grained groundmass. The volcanic equivalent of diorite.

anticline A fold in which the oldest rocks are at the center. In most anticlines the beds dip away from the hinge line.

antigorite See **serpentine**.

aragonite (mineral) The calcium carbonate mineral that is stable only above 4 kbar at 50°C and above 10 kbar at 400°C.

asbestos (mineral) As used herein, a fibrous form of chrysotile serpentine, commercially valuable if the fibers are long.

axial plane The plane or surface that joins the hinge lines of a fold and separates it into two limbs.

B

basalt Dark-grey to black fine-grained volcanic rock, relatively rich in iron and magnesium, and with a silica content of 45%–55%. The volcanic equivalent of gabbro.

"bastite" See p. 7.

Bay mud The unconsolidated and loosely flocculated mud deposited in San Francisco Bay during the Holocene.

bed, bedding A layer (or layers) of sedimentary rock whose boundary surfaces are parallel to the surface on which the original sediment was deposited. The bounding surfaces may be planes of parting or surfaces across which the size or character of the constituent grains change.

biotite See **mica**.

blueschist Schist containing the blue amphiboles: glaucophane or crossite. The characteristic rock of the blueschist facies.

blueschist facies See p. 45.

boudinage The pulling apart, or the necking, of a bed, vein, or dike, to make it appear in cross section like a string of sausages (from boudin, the French word for sausage).

bouyancy The ability to float caused by lesser density.

breccia A rock consisting of angular fragments. Volcanic breccia is breccia whose fragmentation occurred during volcanic eruptions; its fragments must average more than 64 mm in average diameter. The average size of fragments in a sedimentary breccia exceeds 2mm.

"broken formation" See p. 8.

C

calcite (mineral) The calcium carbonate mineral, stable at surface pressures; the primary constituent of limestone and marble. Calcite has three perfect cleavages at (i.e., intersecting at angles of) 102° and 78° which break it into rhombs; it is commonly colorless, transparent, or white; it can be scratched easily with a knife; and it effervesces in cold, dilute hydrochloric acid.

carnelian See p. 43.

chalcedony (mineral) A cryptocrystalline (exceedingly fine grained) form of quartz.

channels, channeling See p. 22

chert A hard, fine-grained sedimentary rock composed almost entirely of chalcedony or opal. Chert is the ultimate result of the consolidation of accumulations of the siliceous shells of radiolarians or diatoms.

chevron fold A fold with plane limbs and a sharp, angular hinge line, having a V or zig-zag pattern in cross section.

chlorite (mineral) A family of sheet silicate minerals, containing magnesium, iron, and aluminum, formed under low-temperature and, generally, low-pressure metamorphic conditions. Chlorite is usually green and has one perfect cleavage.

clay minerals (minerals) A group of fine-grained sheet silicates stable only at low temperatures and pressures; generally formed by the weathering or near-surface hydrothermal alteration of other minerals. Clays are the common minerals of soils and fine-grained sedimentary rocks such as shale. They form dull, white-to-brown earthy masses that are plastic and sticky when wet.

claystone Sedimentary rock consisting almost entirely of clay minerals or of clay-size particles but lacking the closely spaced parting of shale.

cleavage (of a mineral) The tendency to split readily along certain planar directions closely related to crystal structure and crystal form, across which the ions are relatively weakly attracted. Cleavage may be perfect—giving mirror-smooth planes; good; or imperfect—giving rough planar surfaces. A mineral may have one, two, three, or four directions of cleavage, whose intersections show up as striae or minute straight steps on cleavage surfaces.

cobble A stream- or wave-rounded rock or mineral fragment whose average diameter is between 64 and 256 mm (between pebbles and boulders).

coeval Of the same age or period.

colluvium Fragmental material on a hillside derived by downslope movement from rocks upslope or by completely shattering and mixing the bedrock beneath which it is found.

Colma Formation See pp. 9, 51.

conformable (of sedimentary rocks) A sequence of beds that represents deposition uninterrupted by erosion or by tectonic deformation.

conglomerate A sedimentary rock consisting largely of

boulders, cobbles, pebbles, or granules, with interstitial sand; a cemented or consolidated gravel.

consolidation The conversion of loose or soft material to harder material, i.e., of a sediment into a sedimentary rock.

contact The boundary line between the two areas over which two formations, or definable rock bodies, come to the surface; also, the buried surface of contact between these two rock bodies.

continental slope The relatively steep slope between the continent or continental shelf (sea with depth less than 200 m) and the ocean floor (deeper than 2 km).

continental margin The edge or boundary of continental crust.

correlation (in geology) Establishing the temporal equivalence of two rock bodies, fossils, or other geological features.

Cretaceous See Geologic Time Scale.

Crop out (of a rock) To be exposed at the ground surface.

cross-bedding See p. 22.

crossite (mineral) A dark blue sodium-, iron-, and magnesium-bearing amphibole, close in composition to glaucophane.

crust (of the earth) The outermost layer of the solid earth, bounded below by the Mohorovicic Discontinuity, a surface of abrupt change in the velocity of earthquake waves, which is about 10–12 km beneath the oceans and about 35 km beneath the continents.

cylindroidal Folded surfaces, all parts of which have one direction in common.

D

dacite A light-grey volcanic rock with 60%–70% silica and commonly with phenocrysts of plagioclase, K-feldspar, quartz, and biotite. The volcanic equivalent of granodiorite.

deformation (tectonic) The bending, folding, faulting, or other deforming of rocks that is caused by the earth's internal forces and the movement of immensely large rock bodies.

density The mass or quantity of matter per unit volume of a material.

devitrify To crystallize into a mass of fine crystals from a glassy condition.

diabase See p. 4.

diatoms One-celled aquatic plants that secrete a siliceous shell.

diatomaceous Consisting largely of shells of diatoms.

dike A sheetlike or tabular igneous intrusion. A dike in sedimentary rocks cross-cuts bedding. (A sheetlike intrusion in sedimentary rocks parallel to bedding is called a sill.)

diopside (mineral) A pale-green calcium- and magnesium-bearing pyroxene common in metamorphic rocks.

diorite A plutonic rock consisting mainly of the plagioclase feldspar that is richer in sodium than in calcium, together with either hornblende, augite, or biotite. Its silica content is usually 55%–65% (cf. **andesite**).

dip (of an inclined surface) The angle of inclination of the steepest line on the surface; this line is always in a direction perpendicular to the strike (cf. **strike**).

dip-slip Displacement on a fault along the line of dip of the fault surface.

disintegration Mechanical breaking apart of a rock into small fragments.

dome (rhyolite dome) A domelike or bulbous extrusion of rhyolite from a volcanic vent; it takes a domelike form because of the high viscosity of the rhyolite lava.

dune A hill of loose sand, heaped up and frequently caused to migrate by winds or by currents of water.

E

eclogite A dense coarsely crystalline metamorphic rock consisting of the bright green sodium-rich pyroxene called omphacite and the red magnesium-aluminum garnet called pyrope and which has approximately the chemical composition of basalt. Eclogite can be formed only at pressures equivalent to depths within the mantle.

en echelon An arrangement of a series of linear features that are parallel but are progressively offset in the same sense (i.e., either all offset successively to the left or all offset successively to the right).

euhedral (of a mineral grain) Bounded by its own inherent crystal faces and therefore exhibiting its inherent crystal form. The opposite situation is called anhedral.

exotic block A block of rock in melange of strikingly different lithology and probable origin from the matrix and common rocks of the body of melange.

F

facies In sedimentary rocks: The sum of features of a rock that characterizes it as having been deposited in a given environment. In metamorphic rocks: The assemblage of rocks that are considered to have been formed under a given set of conditions of temperatures and pressure. See also p. 45.

fan A depositional form on land or on the seafloor, having the shape of a sector of a very flat circular cone whose apex is at the mouth of the canyon from which the sediment making up the fan came.

fault A fracture in the earth along which opposite sides have moved past each other.

fault creep Slow displacement along a fault, either gradually or by frequent, minute, jerking motions, in contrast to the abrupt larger displacement that produces an earthquake.

feldspar (mineral) The most important group of rock-forming silicate minerals. In their atomic structure, feldspars consist of three-dimensional frameworks of connected silicon-oxygen tetrahedra in which one quarter to one half of the silicon ions are replaced by aluminum and the charge is balanced by the presence of ions of potassium, sodium, or calcium. Feldspars are colorless, white, or pink and may occur as thin, rectangular plates with narrow rectangular cross sections. They have two cleavages, one perfect and one imperfect, at nearly right angles (cf. **plagioclase**).

K-feldspar (mineral) Feldspar that contains potassium.

ferric oxides (mineral) Red, yellow, and brown, usually earthy, oxides of ferric iron, the result of weathering and oxidation of iron-bearing minerals exposed to air and water.

flame-structures See p. 22.

flute casts See. p. 35.

fold A curve or a bend in the bedding of rocks, e.g., an anticline or syncline.

foliation The tendency of a rock, especially a metamorphic rock, to split into thin sheets or flakes, usually along the cleavage direction of platy or fibrous minerals.

foraminifera   One-celled marine animals that secrete a perforated shell of calcium carbonate (lime); foraminifera are important microscopic fossils for dating rocks ranging in age from Ordovician to Holocene.

fossil   An organic trace, including skeletal remains, shells, and impressions, buried by natural processes and preserved in the rocks.

friable   Easily crumbled or crushed to a powder.

G

gabbro   See p. 4.

garnet (mineral)   A group of common metamorphic minerals with isolated silicon-oxygen tetrahedra in their structure. They occur commonly as well-formed regular dodecahedrons (12-sided solids) and are generally red or brown. They are hard and are denser than other minerals of their composition.

glaciation   The process of covering large land areas with thick ice masses (glaciers) accumulated from the compaction of many years of winter snow that persists through the summer; also the processes and results of erosion by glaciers.

glaucophane (mineral)   A blue, sodium-bearing amphibole characteristic of the blueschist facies; stable only at relatively low temperatures and formed only under high pressure. Occurs as silky blue fibers and as deep-blue to black needles.

glaucophane schist   Schist containing an abundance of glaucophane.

gneissic banding   Parallel arrangement of minerals and their segregation into layers as a result of metamorphism, but lacking the easy splitting of a schist.

gouge (fault gouge)   A rock flour formed by crushing during faulting.

grade, grading, graded bedding   See p. 22.

granite   A coarse-grained plutonic rock that contains more than 20% quartz and in which more than one third of the feldspar is K-feldspar. Biotite is the common dark mineral (cf. **rhyolite**).

granitic rocks   The group of coarse-grained plutonic rocks that consist predominantly of feldspar.

granules   Detrital grains between 2 and 4 mm in average diameter, intermediate between sand and pebbles.

Great Valley Sequence   See p. 56.

greenstone   See p. 7.

greywacke   See p. 7.

grossularite (mineral)   The calcium-aluminum garnet resulting from the metamorphism of calcium-rich rocks such as marble and basalt.

groundmass   The fine-grained or glassy body of a porphyritic igneous rock in which are set phenocrysts.

H

hand lens   A pocket magnifying glass for examining objects under magnifications of from 5 to 20 times.

hinge line, hinge   The line of sharpest bending (or the central line of the most sharply curved part) of a fold or a folded bed.

Holocene   See Geologic Time Scale.

hydration   The combination of water molecules into the crystal structure of a mineral.

hydrothermal alteration   The alteration of minerals in a rock to other minerals by the action of hot water (which may contain dissolved materials) from deep within the earth.

hydromagnesite (mineral)   A hydrated magnesium carbonate occuring as dull-white earthy crusts or small ball-like masses; a product of low-temperature hydrothermal alteration of serpentine.

I

igneous rock   A rock that has consolidated from a partly or wholly molten state, i.e., from a lava or magma. Includes volcanic and plutonic rocks.

inclusion   A fragment of an older rock incorporated into another rock, especially a fragment of older rock incorporated into an igneous rock.

interbedded   Occuring as alternating beds.

intercalate   To insert between or within.

interstices   Small or narrow void spaces, e.g., between sand grains.

interstitial   Occuring in interstices.

intrusion   The invasion of magma into or between solid rock, and the body of rock resulting from such an invasion.

ion   An electrically charged atom. Positive ions have lost electrons and are attracted to negative ions, which have gained electrons.

J

jadeite (mineral)   A very dense, sodium-bearing pyroxene formed by the alteration under high pressure of sodium-rich plagioclase. A characteristic mineral of the blueschist facies.

jasper (mineral)   A brightly colored tough opaque form of chalcedony with a conchoidal (shell-like) fracture, sometimes used as a semiprecious stone.

Jurassic   See Geologic Time Scale.

K

Keel   A downward, ridgelike or riblike projection, such as a ship's keel.

kilobar (kbar)   One thousand bars (1 bar = 10^5 N m^{-2}, approximately the pressure of the atmosphere at sea level). One kbar is approximately the gravitational pressure produced by 3.6 km of rock with an average density of 2.8×10^3 kg m^{-3}.

kink (kink fold)   A thin, diagonal belt of sharply bent bedding or schistosity caused by compression parallel to the bedding or schistosity. The inclination of the belt and the angle of bending are such that no change in volume occurs on kinking. Once formed, kinks can broaden gradually until all the bedding or schistosity has the new directions. Folds so produced are called kink folds.

L

laminae   Beds less than 1 cm thick.

laminated   Very thinly bedded.

lava   Molten rock matter extruded onto the surface of the earth; a congealed flow of the same.

lawsonite (mineral)   A hydrated calcium-aluminum silicate with composition of calcium plagioclase plus water; high density, formed under high pressure and stable only below 650°C. Crystals large enough to be seen are pale-blue blades with two good cleavages at right angles. A characteristic mineral of the blueschist facies.

left-lateral   See p. 55.
lenticular   In the form of thin lenses.
limb (of a fold)   One of the two sides of a fold; separated from the other by the hinge line and axial plane.
lineation   A pervasive linear structure or texture within a rock, such as a tendency of most rod-shaped mineral grains to be parallel to each other.
lithologic   Of or pertaining to rocks.
load cast   See p. 22.

M

m   Meters.
magma   See p. 4.
magnesite (mineral)   Magnesium carbonate, which occurs as dull, earthy, or porcellaneous white to grey masses; an alteration of serpentine.
manganese nodules   Small, usually rounded bodies rich in magnanese oxides, abundant on parts of the deep ocean floor.
mantle (of the earth)   The spherical shell of the solid earth between the base of the crust (which see) and a depth of 2900 km. It is solid and thought to consist of high-pressure minerals composed mainly of iron, magnesium, silicon, and oxygen.
massive   Without obvious internal structure such as bedding.
melange   See p. 6.
Mesozoic   See Geologic Time Scale.
metachert   A metamorphic rock derived from chert.
metamorphic rock   A rock formed by the action of high temperatures, high pressures, and hot water solutions on other rocks at depth within the earth's crust.
metamorphism   The process of making a metamorphic rock from a preexisting rock.
mica (mineral)   A group of sheet silicates, common constituents of plutonic and metamorphic rocks, with one perfect cleavage on which they can be split into flexible and elastic flakes. Potassium is an essential constituent. Common micas are muscovite (white mica) and biotite (iron-bearing black mica). Also as detrital grains in sand and sandstone.
m.y.   Million years.
Miocene   See Geologic Time Scale.
mullion   A groove on a fault surface parallel to the line of displacement on the fault.
mudstone   A rock consisting of clay minerals but lacking shaly parting.
muscovite   See **mica**.

N

No entries.

O

olivine (mineral)   A magnesium-iron silicate with isolated silicon-oxygen tetrahedra. The essential constituent of many ultramafic rocks and present as scattered crystals in some basalt and gabbro; possibly a major constituent of the earth's mantle. It has altered, under near-surface conditions, to serpentine. A clear, pale-green, glassy mineral without good cleavage, usually in sugary aggregates. It weathers readily to become coated with iron oxides.
ophiolite   See p. 5.
ophitic   See p. 7.

overturned fold   A fold, one of whose limbs is overturned.

P

paleo   Prefix meaning ancient or of a past geologic time.
parting   A plane across which a crystal will divide, which is not a true cleavage plane; one of many parallel planes along which a rock, such as shale, splits readily.
pelecypods   Mollusks with two hinged, generally symmetrical shells, important late-Paleozoic to Holocene fossils.
phenocryst   A large commonly euhedral crystal embedded in the groundmass of a porphyritic rock.
pillow, pillow basalt   See p. 4, Figure 35.
plagioclase (mineral)   The group of feldspars containing sodium and calcium. All proportions from pure sodium-bearing (albite) to pure calcium-bearing (anorthite) are possible. Plagioclases are the essential minerals of basalt, gabbro, andesite, diorite, dacite, and granodiorite; they are also abundant as detrital grains in greywacke and in some sands. Color: white or grey, glassy when fresh, chalky when weathered. Two cleavages almost at right angles, one perfect (which may show fine striations) and one good. Commonly as thin plates with narrow, rectangular cross sections. Weathers readily to clay minerals.
planar parallelism   Parallel orientation of flat or rodlike minerals so that their major dimensions are parallel to a single planar direction.
plate tectonics   See p. 2.
platelets   Tiny plates or platelike crystals.
Pleistocene   See Geologic Time Scale.
Pliocene   See Geologic Time Scale.
plunge, plunging   The angle of inclination with the horizontal of an inclined line or linear feature measured in the vertical plane containing the line; used as a verb, its direction of downward inclination.
plutonic rock   Igneous rock crystallized from magma within the earth's crust.
polarizing microscope   A microscope with mechanisms for causing all the light to be polarized that passes through the mineral grain or rock slice under study. (Polarized light vibrates in only one plane.)
porphyritic, porphyry   An igneous rock containing phenocrysts embedded in a groundmass.
pseudomorph   A grain of a mineral, which has replaced or is the alteration of another mineral and preserves the external crystal form of the original mineral.
pyrite (mineral)   Iron sulfide, commonly known as fools' gold. Commonly as brass-yellow metallic cubes with striated faces. Usually present in veins with other minerals. Weathers readily to iron oxides and sulfuric acid.
pyroxene (mineral)   The family of silicate minerals whose structure is built on single chains of silicon-oxygen tetrahedra. Pyroxenes are common metamorphic minerals, the dark minerals of basalt and gabbro, and are present in some ultramafic rocks. See **diopside and jadeite.**

Q

quartz (mineral)   Silicon-dioxide, a transparent to translucent hard mineral with a glassy luster and conchoidal fracture, extremely resistant to weathering. An essential constituent of granite, granodiorite, and a phenocryst in rhyolite; a common metamorphic mineral also occurring in veins, sometimes as transparent crystals with three- or six-sided columns and pyramids. The major constituent of most sandstone and some conglomerate and of quartzite.

quartzite A metamorphic rock consisting of interlocking quartz grains, the product of metamorphism of quartz-rich sandstone.

Quarternary See Geologic Time Scale.

R

radiolarians Microscopic, one-celled, floating marine animals whose complex shells and surrounding spicules (fine, needlelike appendages) consist of silica.

recumbent fold A fold whose axial plane is nearly horizontal.

rhyolite A light-colored, fine-grained to glassy volcanic rock containing 70%–75% silica with abundant aluminum, potassium, and sodium. Phenocrysts of quartz, feldspar, and biotite are common. The volcanic equivalent of granite.

ribbon chert See p. 5.

right-lateral See p. 55.

rodingite See p. 49.

S

sandstone A rock composed of consolidated or cemented sand.

scarp An abrupt steep slope or cliff.

schist A metamorphic rock that splits readily into platy fragments along pervasive surfaces of splitting called schistosity and **foliation**, which are usually along the cleavage of parallel crystals of sheet silicates or amphiboles.

schistose Splitting readily into platy fragments as does a schist.

sedimentary rock A rock composed of fragments of other rocks deposited on the surface of the earth, of material precipitated out of surface waters, or of accumulation of organic remains.

sedimentary structure A feature in a sedimentary rock that indicates its mode of formation or environment of deposition.

seismograph An instrument for recording vibrations within the earth, such as those involved in earthquakes.

serpentine (mineral) A group of hydrated sheet silicates consisting of magnesium, silicon, and oxygen. The serpentine minerals are alterations of olivine and magnesium-bearing pyroxene. The serpentine minerals are antigorite (high temperature) and chrysotile and lizardite (both low temperature) and can only be distinguished from each other by X ray analysis. Also, used herein for the rock consisting almost entirely of serpentine minerals.

shale A sedimentary rock consisting largely of clay minerals, having closely spaced parting parallel to the bedding.

shear zone A thin zone within rocks along which the movement of opposite sides past each other has been accomodated by many small displacements on many planes within the shear zone.

shingle Pebble-sized material on a beach.

silica (mineral) Silicon dioxide (SiO_2), either in a pure form or as a chemical component of minerals and rocks. Quartz is the most abundant pure silica mineral.

silica-carbonate rock See p. 34.

silicate (mineral) A mineral containing silicon and oxygen plus other elements. The fundamental building block of the crystal structure of silicates is the silicon–oxygen (Si–O) tetrahedron, a symmetrical packing of four nega-tively charged oxygen ions with centers approximately 1.4 x 10^{-8}cm apart, and with a positively charged silicon ion in the center. An isolated Si–O tetrahedron has an excess of four electrons and therefore a -4 charge, and the excess negative charge is balanced by positive ions such as magnesium, iron, calcium, potassium, sodium, etc. The net charge on an assembly of Si–O tetrahedra may be reduced by sharing oxygen ions, and the silicate minerals are classified on the basis of how many and how the oxygen ions are shared. Thus the crystal structure may contain isolated tetrahedra, tetrahedra in pairs, rings of tetrahedra, single or double chains, sheets, and three-dimensional networks. Aluminum may substitute for silicon in the tetrahedron, giving rise to different negative charges. Water molecules (H_2O) and hydroxl (OH) ions may be present in the structure. Oxygen and silicon constitute 74% by weight and 95% by volume of the earth's crust; hence silicates and silica minerals such as quartz are by far the most numerous and important rock-forming minerals, and most minerals seen on these trips are silicates.

silicon-oxygen tetrahedron See **silicate**.

siliceous Consisting largely of silica, especially free silica such as quartz or opal.

siltstone A sedimentary rock consisting largely of particles of silt size (between 1/256 and 1/16 mm).

slickenside A fault surface polished and striated by fault movement. Striations are parallel to the last direction of movement. Slickenside surfaces are commonly stepped or toothed. The direction parallel to the striations along which the surface feels smooth is the direction the near (or missing), side of the fault moved with respect to the side on which the slickenside is still preserved.

"slickentite" See p. 7.

spreading zone See p. 2.

stilpnomelane (mineral) A brown micalike sheet silicate, rich in iron, formed at somewhat lower metamorphic temperatures than biotite. It can be distinguished from biotite only be examination with a polarizing microscope or by X-ray analysis.

strand (of a fault) Large complex faults have had movement at different times on any one of a number of nearly parallel intersecting fracture surfaces within the fault zone. Any such surface is a strand of the fault.

strike The direction of horizontal lines on an inclined surface. If the surface is curved, the strike is in the direction of horizontal lines on a tangent plane at the point in question. Strike and dip completely specify the orientation of a plane in relation to the earth (cf. **dip**).

Strike-slip See p. 55.

syncline A fold whose youngest rocks are in the center. Generally the beds dip toward the hinge line of a syncline.

T

talus An accumulation of rock fragments at the base of a rock outcrop or cliff, or more precisely, the slope on such an accumulation.

tectonics See p. 2.

terrane A region or area within which a specific group of rocks predominates.

terrestrial Related to, or deposited on, land, as opposed to within or beneath the seas.

thermal spring Hot spring.

thrust A flat or near-horizontal fault on which the upper plate appears to have pushed upward in relation to the lower plate. Displacement on a thrust shortened the crust, or part of it.

tourmaline (mineral) A hard, boron-bearing ring silicate crystallizing as columns with three curved sides, hence with shield-shaped cross sections, resistant to weathering and erosion. The common variety is iron bearing and black, but rare mangenese- and lithium-bearing varieties may be transparent and pink, purple, green, or blue, and are gem stones. Common in some granitic rocks, and as an interstitial material cementing some other rocks.

tourmalinized Impregnated or cemented by interstitial tourmaline.

transform fault See p. 2.

trend The direction of a line.

tuff A fine-grained deposit or rock composed of material ejected from a volcanic vent; same as volcanic ash.

turbidite See p. 5.

turbidity current See p. 6.

U

ultramafic, ultramafic rock See p. 4.

unconformity A surface of erosion between two sequences of sedimentary or volcanic rocks, which may also, in some cases, contain evidence of tectonic deformation of the older sequence before deposition of the younger.

V

Valanginian See Geologic Time Scale.

variole See p. 55.

vein A thin, sheetlike body consisting of one or a few minerals deposited by hydrothermal solutions in a crack in other rocks.

veinlet A small or microscopic vein.

vesuvianite (mineral) Also called idocrase, a calcium-, magnesium-, and iron-bearing aluminum silicate related in structure to garnet; commonly formed by the metamorphism of calcium- or magnesium-bearing rocks.

volcanic rock An igneous rock consisting of material erupted onto the earth's surface, either as lava or as volcanic breccia or tuff.

volcanic ash See tuff.

W

wavelength (of folds) The distance between two anticlinal (or two synclinal) hinge lines; measured in a straight line parallel to the regional bedding (by analogy with waves on water).

weathering The alteration of a rock as a result of conditions at the earth's surface, usually involving reaction with water, atmospheric gases, and organic products.

Wisconsinan The time interval of the latest great glaciation in North America, thought to be between about 10,000 and 100,000 years ago.

Z

zeolite (mineral) A group of silicates consisting of an open framework of silicon-oxygen tetrahedra through which calcium and sodium may be easily exchanged. They are products of hydrothermal deposition and alteration, very low-temperature metamorphism, and of weathering under alkaline conditions. They are relatively soft, white-to-transparent, and in fibrous or earthy masses.

Selected Bibliography

In San Francisco, publications of the California Division of Mines and Geology (*) that are still in print are for sale at the Ferry Building; those of the U.S. Geological Survey (+), at the Public Inquiries Office, Custom House: See Trip 4 for locations.

Alvarez, W. D., V. Kent, I. Primola-Silva, R. A. Schweikert, and R. A. Larson, Franciscan complex limestone deposited at 17° south paleolatitude, *Geol. Soc. Am. Bull., 91*, Part 1, 476–484, 1980.

*Armstrong, C. G., and K. Gallagher, Fossils from the Franciscan assemblage, Alcatraz Island, *Calif. Geol., 30*(6), 134–135, 1977.

Atwater, T., Implications of plate tectonics for the Cenozoic tectonic evolution of western North America, *Geol. Soc. Am. Bull., 81*(12), 3513–3536, 1970.

*Bailey, E. H., W. P. Irwin, and D. L. Jones, Franciscan and related rocks and their significance in the geology of western California, *Calif. Div. Mines Geol. Bull., 183*, 177 pp., 1964,

*Bedrossian, T. L., Geology of the Marin headlands, *Calif. Geol., 27*(4), 75–86, 1974.

Berggren, W. A., and J. A. Van Couvering, The Late Neogene: Biostratigraphy, geochronology, and paleoclimatology of the last 15 million years in marine and continental sequences, *Paleogeogr. Paleoclimatol. Paleoecol., 16*(1 and 2), ix–xi and 1–126, 1974.

Bird, J. M. (Ed.), *Plate Tectonics*, 2nd ed., Selected papers from publications of the American Geophysical Union, 986 pp., AGU, Washington, D. C. 1980.

Blake, M. C., Jr., and D. L. Jones, Origin of Franciscan melanges in northern California, *Spec. Publ. 19*, pp. 345–357, Soc. Econ. Paleontol. Miner., Tulsa, Okla., 1974.

Blake, M. C., Jr., and D. L. Jones, Allochthonous terranes in northern California?—A reinterpretation, in *Mesozoic Paleogeography of the Western United States*, edited by D. G. Howell and C. A. McDougall, pp. 397–400, Pacific Section, Society of Economic Paleontologists and Mineralogists, Los Angeles, Calif., 1978.

+Blake, M. C., Jr., W. P. Irwin, and R. G. Coleman, Upside-down metamorphic zonation, blueschist facies, along a regional thrust in California and Oregon, *Geol. Surv. Prof. Pap. (U.S.), 575-C*(C1–C9), 1967.

+Blake, M. C., Jr., J. A. Bartow, V. A. Frizzell, Jr., J. Schlocker, D. Sorg, C. M. Wentworth, and R. H. Wright, Preliminary geologic map of Marin and San Francisco counties and parts of Alameda, Contra Costa, and Sonoma counties, California, scale 1:62,500, *Misc. Field Stud. Map MF-574*, 2 sheets, U.S. Geol. Surv., Reston, Va., 1974.

Bloxam, T. W., Jadeite rocks and glaucophane schists from Angel Island, San Francisco Bay, California, *Am. J. Sci., 258*, 555–573, 1960.

+Bonilla, M. G., Preliminary geologic map of San Francisco south quadrangle and part of Hunter's point quadrangle, California, scale 1:24,000, *Misc. Field Stud. Map MF-311*, 2 sheets, U.S. Geol. Surv., Reston, Va., 1971.

Cogan, W. M., Mechanics of the Lone Mountain landslides, San Francisco, California, *Calif. State Mineral. Rep. 32*, chap. 4, pp. 459–474, 1936.

+Coleman, R. G., Composition of jadeitic pyroxene from the California metagreywackes, *Geol. Surv. Prof. Pap. (U.S.), 525-C*, C25–C34, 1965.

Coleman, R. G., Petrologic and geophysical nature of serpentinites, *Geol. Soc. Am. Bull., 82*(4), 898–918, 1971.

Coleman., R. G., *Ophiolites: Ancient Oceanic Lithospheres?*, Miner. Rocks Ser., vol. 12, edited by P. J. Wyllie, 229 pp., Springer-Verlag, New York, 1977.

Coleman, R. G., and D. E. Lee, Metamorphic aragonite in the glaucophane schists of Cazadero, California, *Am. J. Sci., 260*, 577–595, 1962.

Cox, A. (Ed.), *Plate Tectonics and Geomagnetic Reversals*, 702 pp., W. H. Freeman, San Francisco, 1973.

Davis, E. F., The Franciscan sandstone, *Univ. Calif. Publ. Geol. Sci., 11*(1), 1–44, 1918.

Deer, W. A., R. A. Howie, and J. Zussman, *An Introduction to the Rock-forming Minerals*, 528 pp., Longmans Group, London. 1966.

Delehanty, R., *San Francisco, Walks and Tours in the Golden Gate City*, 340 pp., Dial Press, New York, 1980.

Dewey, J. F., Plate tectonics, *Sci. Am., 226* (5), 56–68, 1972.

Dickinson, W. R., Relations of andesites, granites, and derivative sandstones to arc-trench tectonics, *Rev. Geophys. Space Phys., 8*(4), 813–860, 1970.

Ernst, W. G. (Ed.), *The Geotectonic Development of California, Rubey Volume 1*, 706 pp., Prentice-Hall, Englewood Cliffs, N. J., 1981. (See especially articles by W. R. Dickinson, pp. 1–28; M. C. Blake and D. L. Jones, pp. 306–328; B. M. Page, pp. 329–417, C. A. Hopson et al., pp. 418–510; and C. R. Allen, pp. 511–534.)

Fornari, D. J., A. Malahoff, and B. C. Heezen, Volcanic structure of the crest of the Puna Ridge, Hawaii: Geophysical implications of submarine volcanic terrain, *Geol. Soc. Am. Bull., 89* (4), 605–616, 1978.

Greenly, E., The geology of Anglesey, 2 vols., *Mem. Geol. Surv. G. B.*, 1919.

Hall, C. A., Geology and Paleontology of the Pleasanton area, Alameda and Contra Costa counties, California, *Univ. Calif. Publ. Geol. Sci. 34* (1), 89 pp., 1958.

Hamilton, W., Mesozoic California and the underflow of the Pacific mantle, *Geol. Soc. Am. Bull., 580* (12), 2409–2429, 1969.

Hsü, K. J., Franciscan melanges as a model for eugeosynclinal sedimentation and underthrusting tectonics, *J. Geophys. Res., 76*(5), 1162–1170, 1971.

Hsü, K. J., and R. Ohrbom, Melanges of the San Francisco Peninsula—Geologic reinterpretation of the type Franciscan, *Am. Assoc. Petrol. Geol. Bull., 53*(7), 1348–1367, 1969.

Ingersoll, R. V., Evolution of the Late Cretaceous forearc basin, northern and central California, *Geol. Soc. Am. Bull., 90*, (9), Part 1, 813–826, 1979.

*Jennings, C. W., Geologic map of California, scale 1:750,000, *Data Map 2*, Calif. Div. Mines Geol., Sacramento, Calif., 1977.

Lawson, A. C. (Chairman), The California Earthquake of April 18, 1906, report (2 vols. plus atlas), 451 + 192 pp., State Earthquake Comm., Carnegie Institution, Washington, D. C., 1908.

Maxwell, J. C., Anatomy of an orogen, *Geol. Soc. Am. Bull., 85*, (8), 1195–1204, 1974.

Middleton, G. V., A. R. Bouma, M. A. Hampton, C. D. Hollister, V. D. Kulm, E. Mutti, C. H. Nelson, and R. G. Walker, Turbidites and deep-water sedimentation,

Short course, lecture notes, Anaheim, Calif., May 12, 1973, Soc. Econ. Paleontol. Mineral., Pac. Sect., Los Angeles, Calif., 1973.

Moore, J. G., Mechanism of formation of pillow lava, *Am. Sci.*, *63*(3), 269–277, 1975.

Moore, J. G., R. L. Phillips, R. W. Griggs, D. W. Peterson, and D. A. Swanson, Flow of lava into the sea, 1969–1971, Kilauea volcano, Hawaii, *Geol. Soc. Am. Bull.*, *82*(2), 537–546, 1973.

Murchey, B. L., Significance of chert age determination in the Marin headlands, California (abstract), *Geol. Soc. Am. Abstr. with Programs*, *12*(3), 144, 1980.

Nason, R. D., Investigation of fault creep slippage in northern and central, California, Ph. D. thesis, Univ. Calif., San Diego (University Microfilms, #12,785, Ann Arbor, Mich.), 1971.

Olmsted, R., T. H. Watkins, M. Baer, et al., *Here Today, San Francisco's Architectural Heritage*, 337 pp., Chronicle Books, Inc., San Francisco, 1968.

Olwell, C., and J. L. Waldheim, *A Gift to the Street*, 196 pp., Antelope Island Press, San Francisco, 1976.

Passagno, E. A., Jr., Age and geologic significance of radiolarian cherts in the California coast ranges, *Geology*, *1*(4), 153–156, 1973.

Passagno, E. A., Jr., Upper Jurassic radiolaria and radiolarian biostratigraphy of the California coast ranges, *Micropaleontology*, *23*(1), 56–113, 1977.

Press, F., and R. Siever, *Earth*, 3rd ed., 613 pp., W. H. Freeman, San Francisco, 1982.

+Radbruch, D. H., Areal and engineering geology of the Oakland east quadrangle, California, scale 1:24,000, *Geol. Quadrangle Map GQ-769*, U. S. Geol. Surv., Reston, Va., 1969.

+Radbruch-Hall, D. H., Map showing recently active breaks along the Hayward fault zone and the southern part of the Calaveras fault zone, California, scale 1:24,000, *Misc. Invest. Map I-813*, 2 sheets, U. S. Geol. Surv., Reston, Va., 1974.

Ransome, F. L., The eruptive rocks of Point Bonita, *Univ. Calif. Berkeley Publ. Geol. Sci.*, *1*(3), 71–114, 1893.

Ransome, F. L., Geology of Angel Island, *Univ. Calif. Berkeley Publ. Geol. Sci.*, *1*(7), 193–235, 1894.

Ransome, F. L., On lawsonite, a new rock-forming mineral from the Tiburon Peninsula, Marin County, California, *Univ. Calif. Berkeley Publ. Geol. Sci.*, *1*, 301–312, 1895.

Rice, S. J., Tourmalinized Franciscan sediments at Mt. Tamalpais, Marin County, California (abstract), *Geol. Soc. Am. Bull.*, *71*, 2073, 1960.

Robinson, G. D., Geology of the Hayward Quadrangle, California, scale 1:24,000 with sections and text, *Geol. Quandrangle Map GQ 88*, U.S. Geol. Surv., Washington, D. C., 1956.

+Schlocker, J., Rodingite from Angel Island, San Francisco Bay, *Geol. Surv. Prof. Pap. (U.S.)*, *400-B*(B311–312), 1960.

+Schlocker, J., Geology of San Francisco north quadrangle, California, *Geol. Surv. Prof. Pap. (U.S.)*, *782*, 1974

+Schlocker, J., M. C. Bonilla, and D. H. Radbruch, Geology of San Francisco north quandrangle, California, scale 1:24,000, *Misc. Geol. Invest. Map I-272*, U.S. Geol. Surv., Reston, Va., 1958.

*Snetsinger, K. G., Sheared Franciscan rocks at Candlestick Hill, San Francisco County, California, *Calif. Geol.*, *32*(7), 152–156, 1979.

Swanson, S. E., and P. Schiffman, Textural evolution and metamorphism of pillow basalts from the Franciscan complex, western Marin County, California, *Contrib. Mineral. Petrol.*, *69*, 291–299, 1979.

U.S. Geological Survey, Land use and Land Cover, 1976–1977, San Francisco, California, *Map L-164*, scale 1:100,000, U.S. Geol. Surv., Reston, Va., 1981.

Van Eyesang, F. W. B., *Geologic Time Table*, 3rd ed., Elsevier. New York, 1975.

*Wakeley, J. R., The unique beach sand at Rodeo Cove, *Calif. Geol.*, *23*(12), 238–241, 1970.

Waldheim, J. L., and S. B. Woodbridge, *Victoria's Legacy, Tours of San Francisco Bay Area Architecture*, 224 pp., 101 Productions, San Francisco, 1978.

Whiffen, M., *American Architecture Since 1780, A Guide to the Styles*, (paperback ed.), 313 pp., MIT Press, Cambridge, Mass., 1981.

+Wright, R. H., Map showing the distribution of potassium feldspar and fossils in Mesozoic rocks of Marin and San Francisco counties and parts of Alameda, Contra Costa, and Sonoma counties, California, scale 1:250,000, *Misc. Field Stud. Map MF-573*, U.S. Geol. Surv., Reston, Va., 1974.

NOTES

Trip 1. A Streetcar to Subduction

Trip 2. To Fort Mason and Subducted Sandstone

74

NOTES

Trip 3. Baker's Beach and Fort Point: A Trip to Melange
and Serpentine

Trip 4. A Sedentary Survey of the Structure of the City
(With Side Trips Afoot)

NOTES

Trip 5. Marin Headlands: Pillow Basalt and Chert

Trip 6. A Boat Trip to the Blueschist Facies: Angel Island and the
Metamorphosed Franciscan

NOTES

Trip 7. After Subduction Is Over: A BART Trip to a Transform Fault